© Chad Hunt/*New York Times*/Redux

ABOUT THE AUTHOR

DR. SHARON MOALEM is an award-winning neuroge-
neticist and evolutionary biologist, with a PhD in human
physiology. His research brings evolution, genetics, biology, and medicine together to explain how the body works
in new and fascinating ways. He and his work have been
featured on CNN, in the *New York Times*, on *The Daily
Show with Jon Stewart*, on *Today*, and in magazines such
as *New Scientist, Elle,* and Martha Stewart's *Body + Soul.*
Dr. Moalem's first book was the *New York Times* bestseller
Survival of the Sickest. He lives in New York City.

Also by Dr. Sharon Moalem

Survival of the Sickest:
The Surprising Connections Between Disease and Longevity

how sex works

WHY WE LOOK, SMELL, TASTE, FEEL, AND ACT THE WAY WE DO

DR. SHARON MOALEM

HARPER PERENNIAL

NEW YORK • LONDON • TORONTO • SYDNEY • NEW DELHI • AUCKLAND

To my parents

This book contains advice and information relating to health care. It is not intended to replace medical advice and should be used to supplement rather than replace regular care by your doctor. It is recommended that you seek your physician's advice before embarking on any medical program or treatment. All efforts have been made to ensure the accuracy of the information contained in this book as of the date of publication. The publisher and the author disclaim liability for any medical outcomes that may occur as a result of applying the methods suggested in this book.

HARPER ● PERENNIAL

A hardcover edition of this book was published in 2008 by Harper, an imprint of HarperCollins Publishers.

HOW SEX WORKS. Copyright © 2009 by Sharon Moalem. All rights reserved. Printed in the United States of America. No part of this book may be used or reproduced in any manner whatsoever without written permission except in the case of brief quotations embodied in critical articles and reviews. For information address HarperCollins Publishers, 10 East 53rd Street, New York, NY 10022.

HarperCollins books may be purchased for educational, business, or sales promotional use. For information please write: Special Markets Department, HarperCollins Publishers, 10 East 53rd Street, New York, NY 10022.

FIRST HARPER PERENNIAL EDITION PUBLISHED 2010.

Designed by William Ruoto

Library of Congress Cataloging-in-Publication Data is available upon request.

ISBN 978-0-06-147966-3

10 11 12 13 14   OV/RRD   10 9 8 7 6 5 4 3 2 1

contents

introduction

We're here to explore human sexuality from beginning to end—what we like and why we like it; how it makes us feel; how it can go wrong; and how human intervention, through cultural traditions, scientific discovery, or both, can divert nature's path—across history, geography, culture, gender, and orientation . . . how sex works.

Along the way, we're going to look at nature's silent hand in the development of human sexuality. Over and over you'll be surprised to discover the evolutionary influence on, well, just about everything: the physical characteristics we find attractive and the personality types we're drawn to; the sexual acts that give us pleasure and why; even fidelity, infidelity, chastity, and promiscuity.

The truth is, evolution and sex are joined at the hip; they've been intimately involved for eons. Evolution has two overriding concerns for every single species—reproduction and survival. And it's constantly working to improve the odds of both, through what amounts to a grand genetic game of trial and error. If a new trait gives its owner an advantage at surviving or reproducing, then that trait's going to spread throughout the gene pool of the species as those that possess it survive longer and reproduce more often, passing the trait on to their offspring. And the flip side, of course, is that a trait that makes it harder for an individual to survive or reproduce isn't going to last very long, because its owner isn't going to pass it on much, if at all.

What does all this have to do with sex? Well, if the process of evolution is stirring up the gene pool in a constant search for competitive advantage, sex is the mixing bowl.

Actually, sex isn't the only card evolution has to play.

Nature has other options. Asexual reproduction is actually thought to have come first, and lots of organisms still do it that way. In its simplest form, asexual reproduction occurs through processes such as *binary fission*, which is when single-celled creatures like bacteria, divide into two carbon copies of themselves. So, one of the things we're going to examine is why nature steered us toward sex, when it is so much more complicated and time-consuming.

We're going to begin our exploration of human sexuality where most people outwardly begin theirs—at puberty, where hormones and history collide in a biological and emotional process that begins the transformation of girls into women and boys into men. We'll look at the physical hallmarks of adult sexuality—square jaws, round breasts, male height, female hips—and consider how they relate to our biological goals. And we'll examine what happens when changing cultural habits and standards intersect with chemistry and biology. Why has the average age of a girl's first period plummeted over the last 150 years? In the context of growing awareness over the horrific practice of female circumcision, is there any health benefit to be found in male circumcision or should it be likewise banned? Can women, like men, ejaculate? What's the point of having pubic hair, and what happens to pubic lice when you get a Brazilian wax? And why, unlike almost all other mammals, do most human women outlive their fertility by decades?

Then we'll turn to a subject that has fascinated scientists, artists, and pretty much every man or woman who's ever lived—beauty; or, to put it in technical terms, what creates attraction and arousal? We'll look at the truth lurking beneath the clichés: tall, dark, and handsome men; hourglass-shaped women; and the "spring fever" that brings them together. We'll uncover the fascinating truth behind something most women know intuitively, but science is just beginning to take note of—the powerful role of scent in the chemistry of attraction. Just like a real mom, mother nature takes a keen interest in who you bring home; millions of years of biological engineering are at work trying to help you find the right mate and keep them. And then we'll look at how a popular little pill can throw the whole scent system haywire.

The chemistry of arousal leads directly into a discussion of the sexual act itself. What exactly happens when you have an orgasm? And why do humans have them anyway? Is it possible that all those hopeless romantics who think good sex and real love enhance each other aren't so wrong after all? And if they're right, then what about all those unrepentant cynics who are convinced that cheating is natural? Finally, we'll shatter the myth of another male monopoly on sexual function: women can be orgasmic, as we now take for granted, but did you know that they can ejaculate too? How? When? And, most importantly, why?

Then we'll turn back to the key evolutionary question: why sex? From a biological perspective, sex is very expensive, and I'm not talking flowers and fancy dinners. Reproduction through sex uses lots of resources, in terms of energy spent seeking out sexual partners, competing for them, keeping them, and mating—all for something less than guaranteed, reproductively speaking. So why sex?

When an organism reproduces asexually, it basically clones itself—meaning all the parasites that it may have acquired through life. One of the biggest advantages of sexual reproduction is that it allows parents to wipe the biological slate clean and occasionally protect their children from some of the sexual and biological misadventures they themselves may have had. The other advantage is the genetic reassortment mentioned earlier: every time we throw the genetic dice, there's a chance that new traits will develop in our offspring, like an immune system that can outwit existing or emerging viruses or bacteria.

But evolution is all about trade-offs: life on two legs makes us taller, but slower. Sexual reproduction gives a chance to protect our children from inheriting parasites and gives them a chance to develop new traits, but it comes with its own set of liabilities. For starters, sex can be a lot more error-prone, for the simple reason that it takes males and females to pull off successfully. So we'll look at how sexual reproduction can go wrong, which happens a lot more often than you might think. When it does—when a child is born with ambiguous genitals, faulty reproductive wiring, hormonal imbalances, or a combination of these disorders of sexual development—what can be done to help that child? More to the point, what *should* be?

Sexual reproduction obviously requires sexual contact with a member of the opposite sex (excepting modern technology, of course). So what happens when individuals are attracted to others of the same gender? And if homosexuality prevents reproduction, why is it common in so many different cultures around the world—especially in men? For that matter, why is it common in so many species: sheep do it, monkeys do it, even killer whales do it.

One of the biggest "costs" of sex is that it exposes individuals to the possibility of sexually transmitted infections (STIs). As always, no evolutionary development in one creature goes unexploited by another—we'll look at how STIs take advantage of biological parts that actually arose to help protect the next generation from parasites and pathogens. We'll explore how some of these invaders may even subvert the chemistry of desire to aid their own reproduction, by making their carriers more *explorative*. And, of course, we'll look at how to prevent them.

Speaking of prevention, we'll examine the surprising proliferation of natural contraception throughout the plant and animal kingdoms.

One thing to keep in mind: this is a book about how sex works, which means it's a book about how body parts connect to each other and to other bodies. That means every once in a while we're going to take a little anatomical detour into the why and how, in order to enhance our understanding of the complexities of human sexuality.

In that spirit, let's begin at the beginning—anatomically speaking, that is. As we set out to explore how sex works and examine all the fascinating differences between male and female sexuality, keep this in mind: ovaries and testes—and the sexual organs most likely to give us pleasure when stimulated, the penis and the clitoris—started out in the very same place, from the very same parts.

girls just want to have fun

If you're a woman, you almost certainly remember the first time you got your period. The first menstruation, called menarche, is only one in a series of events that mark the transformation of girls into young women, but it is one that has been loaded with cultural significance throughout human history. For a long time, menarche was thought to coincide with the onset of fertility. Now we know that most girls do not ovulate with menarche; in fact, it can take a year or two after their first menstruation before ovulation even becomes regular.

Something strange is affecting the age of menarche, making parents and researchers take note and wonder alike. The average age of menarche has crashed from a *traditional* seventeen to just twelve, in the evolutionarily brisk span of just 150 years. So what is turning young girls into young women so quickly? There are lots of theories, but no clear answers.

According to the *psychosocial acceleration theory*, the root cause is increased stress, and a few studies have, in fact, found a correlation between increased stress and earlier menarche. Here's the theory: if young girls in our increasingly complex society experience a lot of stressors early on, their bodies take it as an indication that they have been born into stressful times. In earlier eras, stress was usually the result of circumstances that threatened survival, such as conflicts or famine. In those situations, there might be an evolutionary benefit to earlier menarche, because it would give an individual a chance to reproduce faster, perhaps before succumbing to local threats. For most women in the developed world, the source of today's stress is probably not war or famine. But as far as your brain and body are concerned, stress is stress and it produces the same result.

Then there's a theory that menarche can be triggered in girls who spend little time around their biological father and lots of time around unrelated men. A large study involving 1,938 college women, published in 2006 in the *American Journal of Human Biology*, indicated that both the absence of biological fathers and the presence of half brothers and stepbrothers had an impact on earlier menarche. According to this theory, the absence of one's father and the presence of unrelated men signals a young woman that it's time to start looking for a mate. And how is that signal sent? Well, it might be by scent. As we'll discuss in more detail later, many animals receive chemical signals through smell, and there is real evidence that humans do as well.

Another theory that has been garnering considerably more

weight over the last few years revolves around the skyrocketing rates of childhood obesity. One recent study that looked at weight and age of menarche was conducted by Joyce Lee at the University of Michigan. Lee tracked 354 girls from the time they were three until they were twelve. She found that there was a clear link between extra weight and early puberty. In her study, obese girls—twenty-two pounds or more overweight—had an 80 percent chance of developing breasts before they were nine years old and reaching menarche before they turned twelve.

One study suggests, however, that it's not *how much* fat a girl carries, but *where* she carries it that is driving early puberty. According to William Lassek, a researcher at the University of California at Santa Barbara: "What our findings suggest is that menarche is likely to occur when girls have stored a certain minimal amount of fat in the hips and thighs, and that girls who tend to store more fat around the waist—who have abdominal obesity—are likely to have delayed menarche."

Scientists know that sufficient fat stores are key to the onset of menarche. And as Lassek points out, fat located in the lower part of the body is chock full of the omega-3 fatty acids that are so important for fetal brain development. "This fat is protected from everyday use like money deposited in a bank," says Lassek. "You are not allowed to withdraw it until late pregnancy."

Whatever the biological cause, it's clear that many girls are entering sexual maturity long before they reach emotional maturity, especially today when it takes more emotional maturity than ever to navigate a complicated world. "This long period of mismatch is very confusing for young people," says

Peter Gluckman of the University of Auckland, New Zealand. His book, *Mismatch: Why Our World No Longer Fits Our Bodies*, calls for significant changes in education to help bridge the gap. Gluckman believes that the early age of menarche we're seeing today is likely our set point for menstruation—the norm, given good health and nutrition. According to Gluckman, with the advent of agriculture the overall level of nutrition dropped, resulting in a decrease in nutritional health, and increase in the age of menarche. Early menstruation creates a mismatch for some girls. They may be physically ready, but emotionally and intellectually unable to handle the responsibilities of adult sexuality.

Menarche is essentially the culmination of puberty in girls. Puberty itself is the incredible collection of physiological processes that transform children into adults, with sexually mature bodies capable of reproduction. We're constantly uncovering more of the biological nuances associated with the onset of puberty. For example, scientists have recently discovered a protein called *kisspeptin* (named in honor of Hershey Kisses by researchers at the Penn State College of Medicine), which plays an important role as a biological signal in starting both puberty and ovulation.

But even though the physical transformation into a sexually mature human being is more or less on biological autopilot, adult sexuality is anything but just biological. Modern sexuality is the intersection of biology, society, and history.

What we need, what we want, what we like, and how we like it are all shaped by a combination of evolutionary imperatives, cultural training, and individuality. Evolution, of course, has a keen interest in encouraging us to have sex, even if it comes at a significant cost. The guiding imperative for any species is survival. At least for us, and for most vertebrate animals, no sex means no babies, and no babies means extinction. Having an interest in our sex lives clearly has evolutionary advantages. But before we get too far into how sex works, let's begin by looking at how girls become women, how women become sexual, and how those changes affect the way men (and other women) perceive them.

THE BIGGEST OUTWARD manifestation of puberty is the development of secondary sexual characteristics. In girls, this means breasts and rounded hips, usually accompanied by more body hair; especially the growth of hair under the arms and in the pubic area. Within about two to four years of the onset of puberty, most girls have breasts that are nearly mature.

Human breasts are unique among primates: we are the only species in which breasts enlarge at puberty and remain enlarged throughout life. Among apes and our other primate cousins, breasts only swell when a female is nursing. Even then, they don't swell very much; it's often difficult to even notice them underneath primate body hair. But human breasts have two functions: one parental and the other sexual, and both functions appear to have played an evolutionary role.

Human breasts are composed of fat and modified sweat glands, called *mammary glands*. The fat is what makes them noticeable. The mammary glands can produce milk, which is a specially configured mixture that includes carbohydrates, protein, fat, vitamins, minerals, and hormones—exactly what the baby needs. Besides facilitating mother-child bonding, breastfeeding also provides babies with antibodies, which are not found in commercial formula and can provide crucial protection against infections. This is one of the main reasons breast-feeding is considered so important to an infant's health. The ducts of the mammary glands terminate at the nipple, which is surrounded by a modified circle of darker skin called the *areola*. Areolae (plural for areola) contain sebaceous glands called *Montgomery's glands*, which release a small amount of oily liquid to protect the nipples and the areolae; this is especially important for the prevention of sore and cracked nipples during breast-feeding.

What's amazing about milk production, or lactation, is that the composition of the milk actually changes with the age of the infant, matching his or her changing nutritional needs. Most countries today recognize the importance of providing human milk for infants, but a few nations, such as Norway and Sweden, go one step further to make sure that babies receive age-matched milk. They have developed extensive milk donor programs, essentially, "milk banks" that provide human milk for infants whose mothers cannot breast-feed. Donor mothers are screened for diseases that can be transmitted through breast milk, such as HIV and hepatitis B, and then their milk is collected. The donated milk is usually pasteurized and frozen to reduce the chance of contamination. I first came across milk banks during a research trip to Sweden a few years ago and was really surprised at just how passionately doctors and parents believed the milk banks improved the health of infants. There are a few milk banks in the United States, but nowhere near the same scale. Given the millions of dollars we spend every year on labor and delivery floors to ensure healthy babies, you'd think we could invest a small fraction in new techniques to support one of the oldest infant health aids on Earth.

LIKE EVERYTHING ELSE, breast size and shape, as well as areola size and color, can vary widely in humans. The color of the areola is especially variable, ranging from quite dark to very pink in some fair-skinned individuals. Large areolae are important visual cues, creating the illusion of a larger breast. Breast size may send important visual signals about a woman's potential fertility. Anything that appears to enhance breast size may make a woman more attractive; hence, the importance of the areola.

The average size of breasts actually seems to be getting bigger: one recent British report indicated that the average breast had gone from a 34B to a 36C in just ten years. Pregnancy and breastfeeding have a significant, yet somewhat reversible, impact on breast size as the mammary glands expand and then fill with milk. In the average breast of a woman who is not producing milk, the ratio of glandular to fatty tissue is about 1:1; in a lactating woman it's more like 2:1. And areolae often get considerably darker and somewhat larger during pregnancy and can stay that way after delivery, which may help babies find their mother's nipples. During our history, when clothing was more optional and polygamy the norm, larger and darker areolae may have been a badge of fertility, signaling the possibility of past pregnancy to interested onlookers.

So why are the breasts of human females so different from those of all other female primates? Since it's the fatty tissue that gives them their distinctive roundedness, we should look to the fatty tissue for an explanation. All *kinds* of theories have been offered—the fatty tissue protects the mammary glands and keeps milk warm; it provides an anchor, a substitute for maternal fur that other primate babies cling to when feeding; large, round breasts are a signal to males that their owner is fertile and has the biological resources (in reserve) to be a mother. There is also likely a connection to sexual attraction, since we know that breasts, including the nipples, can swell by as much as 25 percent when a woman is aroused.

Zoologist and bestselling author Desmond Morris believes that female breasts are actually a mimic of the buttocks. Among most primates, the male mounts the female from behind; the bright red coloring on some female buttocks around the genitals acts as a sexual signal. Morris theorizes that as humans became upright and bipedal, the optimum sexual position became face-to-face and females evolved twin globular breasts to mimic the twin globular cheeks of the buttocks.

When it comes to breasts, there are *lots* of theories. But it may also be that the best answer is the simplest—the fat stores are like an insurance policy, they're there to provide energy to potentially pregnant or nursing women when food is scarce. Ample breasts may also send a signal, to anyone who might be interested, that a woman has sufficient fat to support having and nursing a baby.

Men also have breasts and nipples, of course, but what you may not know is that they have mammary glands too. Although it's rare, as compared to women, having breast tissue means that men can also get breast cancer. Under most circumstances, male mammary glands are essentially dormant and men do not lactate, but under certain conditions, men's breasts have been known to produce milk. For example, some prostate cancer patients have received female sex hormones as part of their treatment to slow the growth of their cancer, and those hormones have sometimes triggered male lactation. And transsexual men on high doses of estrogen may also respond to nipple stimulation with lactation.

Men experiencing extreme starvation have also been known to lactate. It is thought that starvation triggers prolactin secretion from the anterior pituitary (located at the base of the brain), causing the male mammary glands to produce milk.

Although it has yet to be fully studied (medical ethics thankfully don't easily allow us to deliberately starve men just to test the hypothesis), male lactation may just be an evolved response that allows men to produce milk to feed their babies in times of extreme starvation. It is also not uncommon for newborn boys and girls to produce breast milk for a week or two after they're born. Their infant mammary glands produce milk because their bodies are still flooded with hormones from their mothers, which they were exposed to in utero. These are the very hormones their mothers' bodies produce to fill their own breasts with milk. Lactation in newborns, which is perfectly harmless, is sometimes called *witch's milk*. Myth has it that witches looking to feed their familiars were stealing it from helpless babies.

When it comes to breasts and nipples, two of each is the norm, but this is by no means an ironclad rule. Why two? It's all about litter size: humans tend to have one or two babies at a time, so two breasts, with two nipples, usually does the trick.

But there is at least a 5 percent chance that an extra nipple will occur—in men as well as in women. Former rapper turned actor Mark Wahlberg has a third nipple; so does British singer and talk-show host Lily Allen. Technically, they're called supernumerary or accessory nipples, and they usually occur along the "milk line," which runs from the armpit, through the normal nipple, down through the groin, and ends at the inner thigh. "Usually" is the operative word—they've been documented as far away from the chest as the bottom of the foot!

Supernumerary nipples can range from a patch that looks like a mole to a complete third breast, with nipple, areola, and milk-bearing mammary glands. In 2005, researchers from the UK discovered the Scaramanga gene, aptly named after the villain from the James Bond novel and film, *The Man with the Golden Gun*, who had three nipples. The Scaramanga gene, now called Neuregulin 3, was initially reported to be involved in breast development in mice. A third nipple is not only a physiological curiosity. A case report in the *New England Journal of Medicine* in 2005 described a forty-two-year-old woman with what appeared to be a mass near her breast. A biopsy later revealed that it was an adenocarcinoma, a type of cancer that originates from glandular tissue. In this case the cancer most likely arose from the woman's third nipple.

Nipples in both men and women are filled with many nerve endings that can be a source of sexual pleasure when stimulated, ranging from mild to intense. Some women have such sensitive nipples that they can experience orgasm from nipple stimulation alone.

BREASTS AREN'T THE only way girls' bodies change over the course of puberty. The increased volume of estrogen coursing through their bodies causes their pelvis and hips to widen. It also dramatically alters the relative amount and distribution of body fat, depositing fat on the hips, buttocks, thighs, and *mons veneris* or *mons pubis* (Latin for pubic mound, the pad of fat underneath a woman's pubic hair in the area above her genitals). Before puberty, the average girl has 6 percent more body fat than the average boy her age; by the time puberty is completed, she has almost 50 percent more.

The extra fat stores on the hips, buttocks, and thighs probably serve the same purpose as the fat stores in the breasts. As we've discussed, fat tissue is a form of portable energy storage, and fertile females are likely to need additional energy for pregnancy and nursing, especially if they are migrating long distances. And, of course, the reason for a woman's wider hips and pelvis (without regard to fat accumulation) is pretty straightforward—it makes childbirth more reasonable, by increasing the size of the birth canal.

But here's where it gets interesting: despite the current obsession with supermodel waifs and androgynous shapes, it seems most gentlemen prefer hips. Across cultures and throughout history, the classic hourglass figure—relatively narrow waist, wide hips—is considered the standard for female attractiveness. Researchers at the University of Texas at Austin worked with colleagues in China and India to examine thousands of works of American, British, Chinese, and Indian literature, some dating as far back as the first century A.D.

to the present. And without exception, when waists came up romantically, they were narrow; when hips came up, they were wide; when breasts were discussed, they tended to be large, although there were certainly some exceptions.

In the early seventeenth century, the British poet John Harrington described a beautiful woman this way:

Her skin, and teeth, must be clear, bright, and neat . . .
Large breasts, large hips, large space between the browes,
A narrow mouth, small waste[sic] . . .

Other social research has produced a similar result. In *Why Sex Matters: A Darwinian Look at Human Behavior*, Bobbi Low, a professor at the University of Michigan, writes: "Across all sorts of cultures with quite different specific ideas about beauty, both men and women see as most attractive a female waist-to-hip ratio of about 7/10 to 8/10."

Why?

Well, here's one thing we know for sure: women with hourglass figures are more fertile. In 1996, Harvard researchers Peter Ellison and Susan Lipson linked higher levels of the hormone estradiol at the right time to higher fertility. A 2004 Polish study that included Ellison and Lipson concluded that women with large breasts, narrow waists, and noticeably larger hips had 30 percent higher levels of estradiol overall and mid-cycle, at the time of peak fertility, than other women. Grazyna Jasienska, the leader of the Polish team stated: "If there are 30 percent higher levels, it means they are roughly three times more likely to get pregnant." Not everyone might agree with Dr. Jasienska's conclusions, but if hourglass figures go hand in hand with higher estradiol, and higher estradiol means higher fertility, then women with hourglass figures are more likely to conceive and pass their genes on—which means evolution will favor hourglasses too.

Devendra Singh, the psychologist behind a University of Texas study of romantic literature, has another theory. Medical research today shows that abdominal fat poses a very different level of health risk than hip and buttocks fat. People with large amounts of belly fat—so-called apple shapes—have a higher risk of heart disease, diabetes, and various cancers than those who carry their fat on their buttocks, hips, and thighs—so-called pear shapes. Dr. Singh thinks people may be programmed to prefer narrow waists because they know they're healthier. Of course, it may be social programming, not genetic programming—we may tend to choose thinner partners today because we equate thinness with health. And, of course, in some cultures, where food is historically scarce, a preference for visibly larger women may have developed because a bigger size is a better indicator of good nutrition and thus fecundity (increased level of fertility). Anyone who's ever been to an art museum or paged through an art history book with a section on Renaissance or Baroque art has seen portraits of women who certainly seemed to have had healthy appetites—and the wealth to sate them.

Even more fascinating, a recent study suggests that curvy moms have more clever kids. William Lassek of the University of Pittsburgh and Steven Gaulin of the University of California, Santa Barbara, used data from the National Center for Health Statistics to show that children whose mothers had wide hips and a waist-to-hip ratio of 7 or 8 to 10 routinely scored higher on intelligence tests. It turns out there's a possible explanation—hip fat contains specific fatty acids acquired through the mother's diet that are critical to development of the brain in fetuses.

IN THE HIERARCHY of attraction, emerging research suggests that one physical characteristic trumps all others—symmetry. By symmetry, I mean exactly that—eyes the same shape, dimples on both cheeks, legs the same length, hands the same size—you name it, left and right sides the same. Across the animal kingdom, males and females find the opposite sex more attractive when their left and right sides match, and humans are no exception.

While some people may find a man born with a crooked nose ruggedly handsome and while Marilyn Monroe's beauty mark may be the asymmetrical exception that proves the rule of her otherwise symmetrically beautiful face, in study after study, the more symmetrical a face, the more attractive members of the opposite sex find it. When pairs of body parts don't match exactly—a right foot bigger than a left foot, a grin that curls up only on one side—it's called *fluctuating asymmetry*. More on the sexuality of symmetry later, but for now, let's consider one more interesting connection between symmetry, sexual attraction, and success in the evolutionary endgame—reproduction.

One of the more noticeable examples of fluctuating asymmetry occurs with female breasts. Many women have breasts that are not the same size or exact shape; in fact, it's actually quite common. Having asymmetrical breasts doesn't affect a woman's ability to nurse, but it may affect her overall health. A British study published in the journal *Breast Cancer Research* in 2006 found that women with significantly asymmetrical breasts are at much greater risk of developing breast cancer. According to the authors, "Asymmetrical breasts could prove to be reliable indicators of future breast disease in women, and this factor should be considered in a woman's risk profile." At the same time, women with evenly matched breasts also appear to be significantly more fertile than women with asymmetrical breasts.

So, if men are more attracted to women with symmetrical breasts, they're also more attracted to women who tend to be more fertile and possibly healthier. To be clear, this is about fertility, not lactation. Large breasts and small breasts are equally capable of producing plenty of milk, even if the same woman has one of each. The mammary glands inside the breasts do the work; it's the fat that surrounds those glands that gives the breasts their size and shape.

Breasts are the most prominent female secondary sex characteristic in humans. But as we know from other primates, they don't need to be large and prominent to succeed at their apparent primary purpose, the feeding of infants. And as I've mentioned, the fat stores are likely there to provide backup energy for pregnant and nursing women. But maybe those fat stores serve another role too.

In many species, males and females (and sometimes both) compete for the sexual attention of the other. Different characteristics make individuals of a given species more or less attractive; everything from behavior, like aggression and dominance, to physical characteristics, like the color of an animal's rump, can come into play. Individuals with the more attractive traits are more likely to mate, which means they are more likely to pass their genes on to the next generation, including the genes for whatever traits make them more attractive. The evolutionary process that selects for those sexually appealing traits is called *sexual selection*. Many of these traits are secondary sex characteristics that serve to advertise the individual's wares—for instance, a peacock's tail feathers or a stag's antlers. And just as corporations allocate enormous sums of money to advertise their products, the body has to consume food and spend energy to create these physical advertisements and keep them at peak appearance. Despite their cost in terms of resources or energy to create and maintain, large, symmetrical breasts are very valuable because they give the individual who possesses them a better chance of attracting a good mate. So it may be that, for humans, breasts function as a kind of cost-effective signal, an easy visual shorthand, flashing a possible projection of future fertility. If you have large breasts, you may have the physical resources, in fat stores and energy, to successfully get pregnant and nurture a child. In some ways, a large buttocks can be just as useful.

BEFORE WE PROCEED any further—a quick note on terms. The main female sex organ is commonly referred to as the *vagina*, but technically speaking, the vagina is all on the inside; it's the passage that extends from the outside of the body to the uterus. The *vulva* is the part of the female sexual anatomy that appears on the outside. Female genitals have had many names and nicknames throughout history, of course—you're almost certainly familiar with some of them. From the late sixteenth century to sometime in the eighteenth century the vulva was referred to as the "hey nonny-no." In the nineteenth century it was the "upright grin." In the 1960s, it was affectionately referred to as a "furburger." We're going to stick with proper names, though—vulva on the outside, vagina on the inside.

The vulva itself has multiple components that have distinct, but related, and sometimes overlapping, functions. At the top is the soft area of fatty tissue usually covered with pubic hair called the *mons pubis* or the *mons veneris*, the "mound of Venus," after the Roman goddess of love. A version of the mons also exists in men. This fat pad is no accident: it actually develops at puberty after hormonal stimulation of the area. Why? Cushioning! It acts like a pillow between partners, protecting the pubic bone during intercourse. Without it, the banging of pubic bone on pubic bone could be a seriously bruising distraction. And, in one of the many ways that humans have evolved to encourage reproduction and make sex more fun, our bodies protect the mons. Even after extreme weight loss, the fat pad of the mons is still there. As Dr. Elizabeth G. Stewart notes in her guide, *The V Book*, "even the skinniest Hollywood actress is as comfortably upholstered for intercourse as her size-12-and-up counterparts in the real world."

Around the time of *thelarche*, when a young girl's breasts begin to bud, a girl may notice pubic hair, emerging around the labial lips and moving up to cover the mons. This is called *pubarche*, and lasts for about six to twelve months. Most women typically end up with a *triangle* shaped pubic hair distribution.

Boys also go through pubarche and grow pubic hair, but instead of *triangle* shaped, most men's distribution is *diamond* shaped.

So why do we have pubic hair? Well, one of the possibilities is that it serves as a human perfume factory. Here's how.

Besides the tops of our heads, the two other prominent places that tend to be covered with hair in men and women, after puberty, are our underarms and our genital areas. These areas are home to numerous specialized *apocrine sweat glands*. These glands secrete sweat that contains fats and proteins. When those compounds are broken down by the microorganisms that naturally make their home in the warm environments under our arms and between our legs (and the rest of our bodies in fact), they produce the distinctive odor that is our very own scent signature. And that scent is a key player in the chemistry of attraction.

Of course, today, women are much more likely to ask, "Why *should* I have pubic hair?" than they are to ask, "Why *do* I?" And for millions of them, the answer is a resounding, "I shouldn't." Or at least not very much. When it comes to pubic hair, exact numbers are hard to come by, but a locker room survey would likely reveal that most women under thirty, and a great number over it, shave, trim, shape, or wax their pubic hair.

The Brazilian—a style of bikini wax that leaves women with a trim "landing strip" of hair on the pubis, or none at all in a full Brazilian—is one of the most popular treatments in the United States and Britain. A new study suggests that the cultural preference for trim pubic hair is having dire consequences for a pesky parasite that has been freeloading in our personal perfumeries for thousands of years. That's right: Brazilians are killing *Phthirus pubis*, also known as the "crab louse" or "pubic louse."

Pubic lice are generally transmitted through sexual contact, like other so-called sexually transmitted infections, or STIs. A team of British researchers tracked the incidence of pubic lice, chlamydia, and gonorrhea over a six-year period from 1993 to 2003. What did they find? While the incidence of chlamydia and gonorrhea both climbed over the period, the incidence of pubic lice declined, especially around 2000 when Brazilians became more widely sought after in the United Kingdom. It seems that waxing down under is like deforestation, as far as pubic lice are concerned.

Our era is by no means the first time humans have removed body hair, particularly pubic hair. For example, ancient Egyptians, and some prostitutes in fifteenth- and sixteenth-century Europe, are thought to have been fond of pubic hair removal. And although we can't be sure, they may have done it just to combat pubic lice. Some Europeans actually wore a pubic hair wig, called a *merkin*, to cover up the shave job and hide the fact that they were trying to prevent the acquisition of the pesky parasites. Today we shave for fashion and, arguably, promote hygiene as a result. Five hundred years ago they shaved for hygiene and covered up for fashion as a result. That's progress for you. Of course, fashion isn't the only reason people in some cultures remove pubic hair; millions of Muslim men and women do it for religious reasons.

Incidentally, it's not only sixteenth-century Europeans who find a thick mound of hair down there more attractive than a clean shave. In Korea, as in some other cultures, pubic hair is an important sign of fertility. But many East Asians, including Koreans, have significantly less prominent body hair than people of Caucasian descent. So Korean women who feel they lack sufficient pubic hair may very well participate in the latest South Korean personal fashion trend: pubic hair transplants.

Pubic hair transplants work like hair transplants to the top of one's head; hair follicles are surgically removed (usually from the back of the scalp) and transplanted, in much the way a fully grown tree might be planted in your yard. At a cost of about $2,500, is it worth the money? Apparently yes—and Korea is not the only place where having pubic hair may be important.

A 2006 paper published in *Aesthetic Plastic Surgery* describes a pubic hair transplant that took place in Brazil. This brings us full circle and suggests another possible biological reason for pubic hair: maybe these cultures view abundant pubic hair as a sign of fertility—because that's exactly what it is. Besides being a perfumery, and a signal of sexual maturity in some cultures, pubic hair may signal that, at least from a fertility perspective, you're ready for business. This may even explain the popularity of leaving a "landing strip" in Brazil and other places; the completely bare look appears sexually immature to some. And like many rules, the notion of pubic hair as a sign of fertility finds proof in its counterpoint: some genetic conditions that leave a woman sterile, for instance, androgen insensitivity syndrome (AIS), may also leave her with a less than normal amount of pubic hair.

BEYOND THE VAGINAL opening, or *introitus*, is the vagina. *Vagina* is Latin for "sheath" or "scabbard"—giving you a pretty clear idea what those old Roman linguists thought about the vagina's role in the world. Today, of course, we know that the vagina is a highly flexible multipurpose organ that can stretch, contract, and manage its own interior climate in order to meet the task at hand. It facilitates sexual activity and pleasure, conception, childbirth, and general maintenance of the reproductive system and allows for the passage of menstrual fluid and tissue from inside of the uterus.

The vagina extends from a woman's vulva, where it opens, to the cervix. Like everything else about human anatomy, there's a lot of variation, but the average vagina is between two and a half to three inches long at rest. But, as we just noted, the vagina is very flexible.

When a woman is aroused, the vagina quickly lengthens, to about four inches or so and will continue to lengthen as a woman becomes more aroused. The walls of the vagina will also expand in width—or contract, as the case may be—in order to provide the right fit for just about any penis, whatever its size. The walls of the vagina also produce moisture to provide lubrication during intercourse and make it more pleasurable. The upper vagina also balloons out, allowing for the puddling of semen after a man's ejaculation. And, when a woman is ovulating—when the chance of conception and reproduction is at its highest—the cervix actually secretes a specific type of mucus that resembles raw egg white in textural quality.

When a woman becomes pregnant and delivers her baby naturally, the prior shape-changing of the vagina is nothing compared to what comes next. During delivery, not only does the vagina (often called the birth canal at this time) lengthen; it becomes wide enough to allow the baby's head and body to pass through it, many times wider than it's ever been at any other time.

The cervix is located at the interior end of the vagina. The word *cervix* means "neck," and the cervix is essentially the neck of the *uterus*. If you were to place a finger into the vagina as far back as you could go, you'll feel something like the tip of your nose—that's the tip of the cervix. The cervix acts as a gatekeeper to the uterus, providing a passageway for sperm to enter and menstrual fluid and tissue to exit.

The cervix is also an important part of the vagina's climate control system, secreting different types of mucus at different stages in the reproductive cycle to help regulate the vaginal environment.

Most of the time the passage from the cervix into the uterus is blocked by thick mucus, but in fertile women it undergoes a slight transformation twice a month, during ovulation and during menstruation. Around the day before a woman's body gets ready to ovulate, her ovaries flood her body with estrogen. The increased level of hormones prompts the cervical opening, called the *os* (Latin for mouth), to become more accessible, in preparation for receiving sperm. At the same time, the cervix produces a specialized "fertile" mucus, which is thinner and less acidic (this is the mucus that resembles egg whites), and thus more hospitable to sperm. All this increases the chances that the sperm will find its egg.

During menstruation, the cervical *os* opens even wider, stretching somewhat to allow menstrual material to flow from the uterus and out of a woman's body through the vagina. The painful menstrual cramps millions of women experience worldwide are the result of uterine contractions, which help the uterus shed its *endometrium*, or lining.

Immediately after menstruation, the cervical opening is blocked for several days by cervical mucus that is the opposite of fertile mucus. It is thick and acidic and highly unfriendly to sperm (and usually other uninvited guests such as microbes), and it prevents sperm from entering the cervix and passing into the uterus.

During childbirth the cervix expands dramatically to allow the fetus to leave the uterus and enter the vaginal canal on its way into the world. You've probably heard doctors talk about cervical dilation as a measure of how close a woman is to giving birth. This is what they're talking about, the expansion of the neck of the uterus from closed to a fully dilated ten centimeters, or about four inches.

The cervix has been observed to relax and contract during orgasm, leading some researchers to suggest—prompting much controversy—that it does so in order to suck sperm from the vagina into the uterus. Others dispute this theory, and the scientific jury remains out.

During intercourse, sperm pass through the cervix and into the uterus, seeking a woman's egg. When a sperm and egg first combine, the single-celled organism is called a *zygote*. The zygote rapidly divides into a multicelled organism called a *morula* (named after the Latin *morus*, or "mulberry," which it resembles), and then it becomes a *blastocyst*. When a blastocyst implants into the inner lining of the uterus, or endometrium, it then becomes an *embryo*, and that's when the fetus starts using its mother's resources to fuel its development in earnest.

The uterus is also connected to the *Fallopian*, or uterine, tubes, which are the twin passageways that eggs follow on their way to the uterus and which sperm follow from the uterus in their search for eggs. Typically, conception takes place in one of the Fallopian tubes. If a blastocyst is impatient, it can implant before it reaches the uterus proper. This is called an *ectopic pregnancy*, and can result in a serious medical emergency. Unchecked, an ectopic pregnancy can lead to a rupture of the Fallopian tube (and its associated vasculature), hemorrhage, and even death. Fallopian tubes are not the only place that can house an ectopic pregnancy. Though rare, some women have delivered babies (through cesarean sections) who developed completely outside of the uterus.

In a fertile woman, the endometrium of the uterus goes through a monthly cycle of growth, shedding, and regeneration; together with the ovarian cycle of ovum, or egg, development and, release, this is the menstrual cycle.

As a woman approaches ovulation, the endometrium becomes rich with blood vessels and tissue, in preparation for sustaining an embryo. After ovulation, if there is no implantation, the endometrium sheds (all the tissue that grew in preparation for an embryo dies): this is called *menstruation*. Then the generative process begins again. Of course, if there is implantation, the endometrium doesn't shed—instead, it provides an interface for the placenta to grow into, nourishment for the growing fetus, and a route out for fetal waste.

Despite their role in shepherding eggs to the uterus, the Fallopian tubes are not directly connected to the ovaries. Rather, they open directly into the abdominal cavity, add this to the long list of reproductive marvels. When an egg is released by an ovary, it floats along inside the chamber that holds the intestines, liver, and so forth until it finds its way into the uterus via the Fallopian tube. Because of the open nature of the Fallopian tubes, there is nothing to stop sperm from actually making their way into a women's abdomen. It is thought that these "rogue" sperm are likely to be picked up and killed by the cells of a woman's immune system.

To aid in their role shepherding eggs into the uterus, Fallopian tubes come fitted with *fimbriae*, special fringelike fronds, at their ends. During ovulation, we think that female sexual hormones spur small hair-like cilia on the fimbriae to beat faster. Cilia, in fact, moonlight as matchmakers beating faster in the presence of sperm in frenzied excitement at the prospect of conceptual union. Just outside the Fallopian tubes, of course, are the female gonads, the *ovaries*. Every female human is born with a full complement of eggs—individual eggs mature before ovulation. Which means, by the way, that half of every human being's genetic makeup was actually formed in his or her *grandmother*: by the time your mother was born, she was carrying the egg that, with your father's sperm, turned into you.

ABOUT TWO YEARS after the first outward manifestation of puberty, the first appearance of breasts and pubic hair, girls have their first period, known as *menarche*. In western countries, the average age of menarche is about twelve and a half years. The body fat connection goes both ways, incidentally; just as the burgeoning rates of childhood obesity are linked to earlier puberty, young women with particularly low body fat often have delayed menarche. The body waits until it has sufficient fat stores—

enough to support a pregnancy—before starting the menstrual cycle. Similarly, endurance athletes—like female marathoners, for example—and women who are excessively thin may stop having their periods because their body fat is so low. It's as if their bodies put their menstrual cycles and thus their fertility into hibernation because they recognize they have insufficient resources to properly support a pregnancy.

Cultural treatment of menstruation runs the gamut—from unmentionable curse to celebrated blessing and everything in between. In some places menstrual blood is thought to have magical powers to ward off evil, heal the sick, and guarantee a bountiful harvest. Others believe it can defile religious ceremonies, poison enemies, and ruin a hunt. It all depends where and when you grew up.

The Asante of Ghana have celebrations to honor the life-giving power of menstruating women, while the Kaska (aboriginal Indians from northern Canada) used to relegate them to special huts for the duration of their periods. Some members of the Greek Orthodox Church encourage menstruating women to abstain from taking communion, and Orthodox Jewish law declares a menstruating woman to be *niddah*, prohibiting sexual intercourse until she has completed her period, waited seven days and then immersed herself in the ritual bath called a *mikvah*. The ancient Romans and modern Moroccans—as recently as last century—both believed menstrual blood could cure illnesses or treat wounds, while the Mae Enga of New Guinea used it as a poison (rather ineffectively, of course).

In many developed countries today, of course, menstruation is the subject of much advertising, as anyone who has ever watched daytime television is well aware. In 2004, advertising for sanitary products was estimated to be a $3.02 billion market in the United States. But sanitary products are nothing new. The ancient Egyptians are thought to be the original makers of disposable tampons; they used softened papyrus. In the fifth century B.C. Hippocrates, the Greek physician often called the "father of medicine," wrote of tampons made of lint wrapped around wood. Japanese women used paper tampons and changed them a dozen or so times a day. Women in Africa used various plants and mosses. And the Rungus in Borneo used nothing at all; instead, they sat on platforms of dried moss or bamboo, letting the menstrual fluid flow freely and periodically rinsing themselves and their mats.

According to Nancy Friedman, author of *Everything You Must Know About Tampons*, despite their long and varied pedigree, tampons fell off the approved list in the United States by the 1930s. They were only used by "women [who] belonged to an exclusive margin of society; they tended to be actresses, athletes, or prostitutes—all dubious professions, in the eyes of 'respectable' women." The sanitary product of choice was a pad or napkin.

The first commercial product was Lister's Towels, introduced in 1896 by Johnson & Johnson, but it is thought that it failed because advertising feminine products was deemed unseemly at the time. In the early 1920s, Kimberly-Clark offered Kotex (from COtton and TEXtile), but they sold it to retailers with an ingenious marketing plan that worked wonders to overcome social stigma.

In their book *Kotex, Kleenex, Huggies: Kimberly-Clark and the Consumer Revolution in American Business*, Thomas Heinrich and Bob Batchelor observe:

To make the product available to the woman who was loath to ask a clerk at a drugstore counter to hand her a box of Kotex from the shelf behind him, Kimberly-Clark encouraged merchants to display the product on countertops, enabling the customer to take a box and pay for it with minimal communicative action. Thus Kotex became one of the first self-service items in the history of American retailing. Women took a box and put their money into a container—the clerk was removed from the transaction.

All those early pads weren't nearly as convenient to use as they are today, by the way—until the introduction of Stayfree and New Freedom pads in 1970, all pads and napkins were actually fitted with a belt.

The modern tampon was invented by Dr. Earl Hass who designed a plug with an applicator but couldn't get it to market. He tried unsuccessfully to sell it to sanitary pad makers, including Kimberly-Clark and Johnson & Johnson, eventually selling it in 1933 to a Denver businesswoman named Gertrude Tenderich. She started a company, made the first products at home with her sewing machine, and called it Tampax. The latest product to see an increase in popularity, although it was invented in the 1930s along with tampons, is the *menstrual cup*. It works by collecting the menstrual flow, unlike tampons that absorb it, and is periodically removed and cleaned. For the sake of convenience some manufacturers also produce disposable menstrual cups.

Most women are familiar with the bloating and cramping that often happens in the days leading up to their periods. But menstrual cramps (the technical term is *dysmenorrhea*) aren't the only cramps women experience in their menstrual cycles. About one in five women experience a distinctive pain on one side of their abdomen right in the middle of their menstrual cycles. This is called *mittelschmerz*, from the German for "middle pain"—which is doubly appropriate because it occurs in the middle of the menstrual cycle and you feel it in your middle, or abdomen.

Mittelschmerz isn't pain related to menstruation, though—it's pain from ovulation. Ovulation occurs when a follicle stretches the surface of the ovary and then ruptures, releasing an egg into the abdominal cavity. Remember, the ovaries aren't directly connected to the Fallopian tubes; instead, when everything works right, the fimbriae sweep that egg up and into the tubes. Until very recently, scientists thought that the release of an egg from the ovary happened very quickly, but an accident of surgery has brought that into question. Doctors in Belgium were preparing to conduct a hysterectomy on a forty-five-year-old woman when they noticed she was about to ovulate, and they managed to tape the whole event. Surprisingly, instead of the expected rapid expulsion, the pictures showed ovulation taking place over the course of about fifteen minutes.

Ovaries usually release only one egg at a time (fraternal twins, which occur when two eggs are fertilized at nearly the same time, are the exception to this rule), and usually only one ovary releases an egg per menstrual cycle. That's why *mittelschmerz* occurs somewhat to the left or right of center—on the side that releases the egg. Some women experience it on both sides, though this is rare. It's not entirely clear what causes the pain; possibly, the release of blood or other fluids when the follicle ruptures irritates the peritoneum, the tissue lining the inside of the abdomen. But clearly, *mittelschmerz* is no cause for alarm—and for women trying to get pregnant, it's actually a little bit of painful good news.

By the way, even though ovulation in humans may cause some discomfort, it's nothing compared to what some animals go through. Cats and rabbits, among others, belong to a class of animals called *induced ovulators*—they don't ovulate on a regular cycle, but only when induced to ovulate by sexual intercourse. And how are they induced? In cats, barbs on the male's penis cause pain on withdrawal; this stimulates the release of hormones, which then cause ovulation. So, if you've ever heard a female cat screaming at the end of intercourse, now you know why. There's good reason to wonder whether it's a scream of pleasure or one of agony.

One more thing about menstruation: you've probably heard the idea that women who spend a lot of time together end up having the same cycles—the apocryphal story is about women in a college dorm. Well, there's a lot of conflicting evidence about this. The original study was by Dr. Martha McClintock in 1971 (which was, in fact, a study of women in a college dorm), and it showed that the participants' cycles synchronized over time. Other studies have shown no evidence of menstrual synchrony. If it does exist, it's likely that it occurs through some form of olfactory signal women pick up from one another about their cycles. In fact, there is a documented synchrony effect in rodents, called the *Whitten effect*, in which the females' menstrual cycles become synchronized when exposed to the urine of males.

Then there's the whole connection between menstruation, months, and moonlight. In our culture we follow the Roman calendar, which is a solar calendar based on the position of the earth in its annual orbit around the sun. Other cultures, such as Islamic ones, have used purely lunar calendars. In a lunar calendar, the length of months corresponds to the length of the moon's cycle from full moon to full moon. And more than a few people have been struck by the similarity between the rhythms of the moon and the rhythms of fertility—the moon cycles every 29.5 days, and the average menstrual cycle is said to be 28 days, although there can be a lot of variation, even in the cycle of the same woman. Does the moon control the menstrual cycle? There is some theorizing that the two are somehow connected; it is thought that the pineal gland (located deep within the brain), receives information about environmental light levels through the eyes and optic nerve, which then affects or interacts with our biological clocks. But most scientists discount this idea. As the astronomer George O. Abell wrote in *Science and the Paranormal*:

> The moon's cycle of phases is 29.53 days, while the human female menstrual cycle averages 28 days (although it varies among women and from time to time with individual women); this is hardly even a good coincidence! The corresponding estrus cycles of some other mammals are 28 days for opossums, 11 days for guinea pigs, 16 to 17 days for sheep, 20 to 22 days for sows, 21 days for cows and mares, 24 to 26 days for macaque monkeys, 37 days for chimpanzees, and only 5 days for rats and mice. One could argue, I suppose, that the human female, being more intelligent and perhaps aware of her environment, adapted to a cycle close to that of the moon, while lower animals did not. But then the 28-day

period for the opossum must be a coincidence, and if it is a
coincidence for opossums, why not for humans?

WHEN IT COMES to anatomy, the clitoris is in a class by
itself. Quite literally. It's the only organ in the human body,
in men or women, that has only one function—to make its
owner feel good. The penis has reproductive responsibilities
and elimination system duties. Breasts have parenting priori-
ties. But the clitoris stands alone. And it stands a bit taller than
you might think.

Most people think the clitoris is just a small nub. But
that's not the case at all. Like the penis, the clitoris has a shaft,
called the clitoral body; from the head of the clitoris it extends back about an inch and a half and then divides into legs,
called *crura*, that extend down around two to three inches,
surrounding the vaginal canal—the whole thing looks like a
large wishbone except the unifying shaft is perpendicular to
the legs. The clitoris is made up of erectile tissue—and when
a woman is sexually aroused, the clitoris (like its counterpart,
the penis) fills with blood and becomes erect.

Of course, clitoral stimulation isn't the only way that
women can have an orgasm. Women can have orgasm through
both vaginal and anal stimulation. Women can have orgasm
when just their breasts or nipples are touched. Women, in
fact, can have orgasm just by thinking about whatever turns
them on.

We know that the pudendal nerve transmits orgasm-
producing sensations received from clitoral stimulation to the
brain; it serves the same function in the penis. So how do sen-

sations from the vagina get transmitted to the brain? Through the pelvic nerve—and there are other nerves that convey information from the cervix and uterus. This possibly explains how all those body parts (including the clitoris), when stimulated together, can produce a "blended" orgasm in some women.

There are also well-sourced reports of breast orgasm going back a half-century or more, including the Kinsey report. More recently, in *The Science of Orgasm*, Komisaruk, Beyer-Flores, and Whipple write: "There are documented cases of women who claim they can experience orgasm just by thinking—without physical stimulation. Their bodily reactions of doubling heart rate, blood pressure, pupil diameter, and pain threshold bear out their claim."

So if women can have orgasms *without* genital stimulation, it seems clear that orgasm is something that happens in our minds. Which means, just like when it comes to our sense of smell or taste, it's something we *perceive* with an astonishing level of variability between people—not just an automatic bodily response to sexual stimulation. And, of course, anyone who has ever been "not in the mood" has experienced the mental blanket your mind can throw over what would otherwise be a most stimulating situation. All of which goes to show that orgasm isn't just a simple reflex.

It's a state of mind.

And there's new evidence that a woman's state of mind about her relationship is directly related to the quality of her orgasms.

Researchers in Switzerland and California used functional magnetic resonance imaging (fMRI) to map the brains of women involved in sexual relationships—and they discovered that the more in love a woman was with her partner, the easier, better, and more intense her orgasms were. That's good news for romantics everywhere. And it gets even better. Because new studies of a powerful hormone called *oxytocin* seem to show that, just as stronger love produces better orgasms, more orgasms may contribute to stronger love.

Oxytocin, called the "love hormone" by some, is involved in all kinds of intimacy. Physiologically, it can trigger lactation, labor contractions, and the jolts of a woman's pelvis when she's having an orgasm. It's found in semen, too. It's also a natural painkiller (but not to be confused with the often abused drug oxycodone). And oxytocin goes through the roof during orgasm—up to five times its normal level. The synthetic version of oxytocin, called *pitocin*, or "pit" for short, has been given to millions of women to induce and speed up labor. Pitocin can actually increase pain because of the sustained uterine contractions, rather than the usual phased contractions that occur naturally.

But oxytocin doesn't just have physiological effects; it can behave as a neurotransmitter too, regulating emotions. It's deeply connected to the mother-child bond, and probably the father-child bond (new studies show that oxytocin levels shoot up when a father holds his baby), as well as bonding between sex partners. Dr. Kathleen C. Light from the University of North Carolina at Chapel Hill has found that oxytocin levels climb not only when couples have sex but even when they hold each other, hug, and hold hands. Although it's still experimental, oxytocin is even thought to help with the symptoms usually associated with autism. Initial studies found that the administration of oxytocin improved the ability of adults diagnosed with autism to comprehend speech colored by emotions such as anger and sadness. It's also thought that the "love drug" Ecstasy raises levels of oxytocin through its main ingredient, the synthetic chemical methylenedioxymethamphetamine, or MDMA, which is thought to explain some of the increased feelings of sociability that the drug produces.

If oxytocin improves bonding between couples, and orgasms increase oxytocin levels in people, well, science has caught up to what millions of couples have known for thousands of years—keeping things good in the bedroom can keep things good outside of it as well.

boys to men

A s boys begin to make the journey to becoming men, there's one piece of anatomy that gets the lion's share of human attention—and most of the myth. And the biggest myth of all is that women always want them bigger. But before you pull out the tape measure, let's add some perspective. One thing's for sure—you're not hung like a gorilla. The truth is, when it comes to penis size, humans dwarf gorillas: an adult gorilla's erect penis is only about 1.5 inches long. In fact, humans have the largest penis of any primate—not just relative to body size but in actual size.

On top of that, men may be surprised to know that their partners are likely to think they look just great down there, even if they don't. A 2006 study found that, while only 55 percent of men were satisfied with the size of their penis, 85 percent of women were quite satisfied with the size of their partner's penis. So, if size doesn't seem to matter to most women, does size matter at all? Well, not in the way you might think. Remember, for most women the most sensitive parts of a woman's genitals are on the outside and just inside her body—the clitoris, the vulva (which includes the labia minora), and the first three to four inches of the vagina (which is also surrounded by the clitoral crura). The average erect penis is over five inches long, and additional length may not improve sexual sensation for all women. When researchers have asked women whether they care about penis size in relation to pleasurable intercourse, the results square with what we know about the physical location of the pleasure centers in the female genitals. Women tend to care a lot less about length than men think—and a little bit more about width. Which makes sense, because additional width may mean more stimulation of the clitoris by causing more pull on the nerve-rich labia minora or stimulation of the introitus and anterior part of the vagina (more on this later).

Of course, there's one way in which size definitely matters—condom fit. About ten years ago, I encountered a rather peculiar local phenomenon while working at a Thai nongovernmental agency (NGO) that was involved in HIV prevention. Western men who frequented the red-light district kept reporting problems with Thai condoms that we were providing—they were breaking. Well, guess what? The locally produced Thai condoms were smaller than those in Western countries.

Although the field of cross-cultural sizing has yet to be fully explored, there are some indications that Europeans are generally larger than Asians, and that's true for their erect and flaccid penises too. And, while the difference may not matter from a sexual perspective, it can make all the difference in the world when it comes to safer sex. For AIDS prevention, the difference in condom and penis size may not be just a cultural curiosity—it can be deadly.

A survey conducted in India found that more than half of Indian men had penises that were about an inch shorter than the international standard that is used to manufacture condoms. If you're an average-sized American who's ever tried to buy clothes in Asia, or if you have a small build and try to find American clothes that fit, imagine what it must be like for some men and ill-fitting condoms.

Condoms, of course, are critical weapons in the fight against disease transmission, especially HIV. And if a condom doesn't fit right, there's a higher chance it will come off or break. This has led some experts to call for the production and advertisement of condoms in a much wider variety of sizes so that size differential doesn't keep men from purchasing them.

So it looks as though one penis myth may have some basis in reality—there may be some variation in penis size across ethnic groups—and the few studies that have been done over the last few years suggest that may be true. But, for the most part, even that difference in penis size may be most pronounced and significant in the flaccid state, rather than the fully erect one. This may be the root of some men's anxiety about penis size; most heterosexual men see many more flaccid penises than erect ones, so they don't realize that there's much less difference once erect. Average penises generally range from three to four inches in length when flaccid and from 4.5 to 5.5 inches when erect. Circumferences can range from three to four inches.

Height seems to vary among ethnic groups, but is the

same thing true for penis sizes? A recent 2007 study of three hundred Indian men from Kerala revealed that their average flaccid penis length was significantly shorter than what was previously reported from studies measuring American and Jordanian men, but similar to the lengths reported from Nigeria and Israel. And, in the handful of studies measuring "stretched length," South Indians from Kerala come out behind Americans, Israelis, and Jordanians but significantly ahead of Koreans. But, when actual erect (and not stretched) penis length was studied, the Kerala group was not that much different from the Americans, although still significantly behind the Israelis.

And why the difference among ethnic groups? Nobody is really certain, but I think it may have something to do with ancestral climate. According to this idea, which doesn't perfectly explain all the ethnic differences, the colder the climate of your genetic origin, the smaller your flaccid penis. Why? The better to protect it from cold exposure and frostbite, of course. And sure enough, there's a correlation to shorter fingers and toes in people from colder climes as well.

As to what all this really proves, the jury is still out. First of all, there's a lack of uniform methodology to penis size studies, and this makes comparisons difficult at best. But one of the biggest limitations to all this research is that many of these studies, including the one from southern India, recruit their participants from clinics for *sexual dysfunction*.

A recent study that tried to sample a group of men who were not seeking medical treatment looked at over three thousand young Italian military recruits between the ages of seventeen and nineteen. The researchers must have been surprised to find that 2.5 percent of them had penises that were under 1.5 inches in length when flaccid and less than 2.75 inches when stretched. The study's researchers reported these lengths to be borderline for the clinical classification *micropenis*, a term usually applied to a stretched penis length below about 0.75 inches in a newborn. Finally, when it comes to penis variation, there's the issue of angle. Men with certain vertical angles (specifically penises that bend toward the abdomen) may provide more pleasure for their partners because they can better stimulate the nerve-rich anterior wall of the vagina, the area where the so-called G spot is said to be found.

In fact, many sexual aids or toys are especially designed for increased stimulation of the anterior vaginal wall. These devices have a similar curve or shape to those horizontally angled penises; they also make it easier for a woman to reach the area, especially when she masturbates. Of course, not all women respond the same way to this or any other type of stimulation, so men can relax: there is no perfect angle, just as there is no perfect size.

There is one angle that can cause some problems, though— for the man, not the woman. A penis with a horizontal angle, one that bends sharply to the left or right, can actually make sexual intercourse rather painful for the man.

UNDERSTANDING STANDARD PENIS size is not a vanity exercise; it's a public-health concern, as the example of the poorly fitting Thai condoms demonstrates. This is especially true when men use nonlatex condoms, which are less forgiving of differences in size than latex. Unfortunately, it's not as easy as it seems to get a clear read on penis size across different populations. Scientists have been measuring penises for more than sixty years, but there's still no consensus on what to measure, where to measure, or who should measure. Should you measure penises that are flaccid or erect? From tip of the glans to the base of the shaft—or all the way to the pubic bone? Should the measuring be done by a doctor in his or her clinic or by the subject in the comfort of his home? Must there be a standard temperature?

They say that you can tell how large a man's penis is from his shoe size—which goes to show how little you can rely on what they say when it comes to penis myths. A 2002 study by two British urologists put that one to bed, showing absolutely no correlation between the size of one's sneakers and the length of one's penis. We still need a standard of measurement for a solid scientific understanding of standard penis length and deviation from it, among different ethnic groups. In the kind of consensus that only an informal international committee of unaffiliated parties could devise, most studies now measure something that only exists when it's being measured—a stretched, flaccid penis. From the pelvic bone to the tip of the glans. At room temperature.

Even better, a pharmacologically induced erection by a professional. That's one everybody can agree on—self-reported results tend to be mysteriously longer.

The normal male penis, whatever its size, is composed of two main parts: the glans, or head, and the shaft. The *glans* is homologous to the head of the clitoris, and like the clitoris, it is jam-packed (although on a per square foot basis not as generously) with highly sensitive nerve endings and quickly responds to stimulation.

The *shaft* is actually composed of the *corpora cavernosa*, twin tube-like compartments that form most of the penis shaft; underneath them, along the center of the shaft's bottom runs the *corpus spongiosum*, which contains the urethra and at the end of the penis becomes the glans. The small opening or slit at the tip of the glans is a multipurpose exit for both semen and urine. In men, but not in women, the urethra has a role in the reproductive system, carrying semen and the sperm within it on the final leg of their journey when a man ejaculates. All three penile tubes, the two *corpora cavernosa* and the *corpus spongiosum*, are made of erectile tissue—tissue that expands when stimulated as blood rushes in and fills its spongy cavities.

That's exactly what an erection is—a rush of blood into the penis. When a man is sexually aroused, nitric oxide (NO) is released in the genital area. Nitric oxide is a vasodilator—it can make blood vessels dilate wider. This produces a flood of blood into the penis, almost all of it into the twin tubes of the corpora cavernosa. These swell and harden, causing the penis to expand in length and diameter and become erect, enabling the penis to penetrate the vagina in intercourse. The corpus spongiosum has a different role to play. It also swells, but to a much lesser degree. It remains softer and pliant in order to keep the urethra open. Otherwise, there would be no way for sperm to get out of the penis. All that work to get a penis inside a vagina would be for naught, reproductively speaking.

So that's what an erection in men is all about—blood and spongy tissue. The one thing that it isn't is bone. "Boner" is a big-time misnomer—at least, that's what you'd think if you only studied humans. Actually, most mammals do have a penis bone, called a *baculum* (and a clitoris bone too). Humans are one of just a few exceptions to the rule; even our closest genetic cousins, chimpanzees, have a penis bone, as do all other primates, cats, dogs, bears, and whales. In Alaska, the baculums of sea mammals, like those of seals and walruses, are called *oosiks*. They are polished and used as knife handles and even sold as souvenirs.

But men definitely shouldn't fret about their boneless member—it helps make sex fun. When it comes to mating, animals have a need for speed because that's a time when they're vulnerable to predators. The baculum makes mating fast: since the penis doesn't need to form an erection to penetrate, there's no need for foreplay, just a quick in and out in thirty to sixty seconds, sometimes even less. Humans' fluid-propelled system takes some time to work up and some stimulation to keep it going; this also allows sex to last longer (but not always). So add the hydraulic penis to the big brain and opposable thumb on the list of things that separate us from most of the animal pack.

EVEN THOUGH HUMAN penises have no bones, they do have something in common with bones, and it's not a happy coincidence. Like bones, human penises can break.

That's right, you can break your penis.

Medically it's called a "penile fracture," and thankfully it's not that common. Here's how it can happen.

The corpora cavernosa are surrounded by a thick layer of tissue called the *tunica albuginea,* which provides the resistance against which the swelling blood within the corpus cavernosum expands during an erection. In cases where an erect penis is bent at an angle that the tissue cannot sustain, the tunica albuginea can rupture. What is the most common way that happens? Someone about to engage in passionate intercourse misses his target and slams into the pubic bone of his partner. And when that happens, he *knows* it. There can be a loud cracking or popping sound, and serious pain. The penis will usually bruise rapidly as blood floods and leaks into the surrounding tissue. I will never forget the first time I came across a penile fracture. The patient had tried to masturbate with a metal vacuum cleaner tube—while it was hooked up and turned on.

In particularly bad breaks, the urethra can even be damaged; this can permanently impair the ability to urinate and to inseminate through intercourse. Usually, however, there is no impact on fertility, because a penis break doesn't involve the testicles, which is where sperm are stored; it's the possibility of natural insemination that's at risk. One of the more common ways for a penile fracture to occur is in heterosexual intercourse, during the down stroke with a semierect penis, with the woman on top.

A seriously broken penis can be repaired through surgery, but it needs to be done quickly. If it ever happens to you, put your penis on ice (a bag of frozen vegetables will do) and get yourself to the emergency room and ask for a good urologist pronto.

In humans, a broken penis is rare, and in the unlucky circumstance that it occurs, it can be fixed. Male honeybees aren't so lucky.

Early in college, before turning my attention to the study of human disease and sexuality, I became interested in insect sexuality while researching how honeybees deal with microbial and parasitic infections. As a result, I came away with a new understanding of just how interrelated processes throughout nature are. In the case of male honeybees, not only do their penises break off when they mate with a queen bee; they die within hours of their first and only sexual encounter. I've always thought that honeybee society exemplified the idea of the disposable male, a society run entirely by females whose male members' only purpose is sexual reproduction. Here is the short, sweet tale of the sex life of a honeybee.

Depending on the time of year, beehives can have a few hundred male drones and thousands of female workers, but only one queen. Only the queen can lay fertilized eggs. Unfertilized eggs become more male drones; fertilized eggs become more female workers. And workers don't have it much better than drones: they don't mate at all; they can only sting once; and when they sting, they die.

The queen is a bee of a very different bonnet. She can sting over and over again and she can mate many times with different males. When the mood strikes a young queen, possibly soon after she is born, she'll leave the hive for a maiden voyage on a spectacular search for sex. (Queens, by the way, are created when workers feed larvae a special diet consisting of a nutritious secretion called royal jelly. All larvae are fed *some* royal jelly, but incipient queens are fed *only* royal jelly.)

As she flies, she casts a trail of pheromones that can bring thousands of males congregating around her, looking for their chance to inseminate her. Mating is in midair; it's quick, but comes with a bang (you can often even hear an audible "pop" as the male drone flips out his penis) for the bee in question. In order for a male to get lucky, he must be near the queen when she opens her sting chamber, making herself available for mating. This may sound rather straightforward, but stop and think for a moment about the spatial orientation, mechanics, physiology, and physics required to bring a queen and male bee together in midair and midflight at heights estimated at more than one hundred feet. The lucky male that manages the acrobatic feat of finding a queen midair rapidly mounts her, pops out his endophallus (the technical name for his penis), and inserts it into her sting chamber (a sort of multipurpose vagina).

Dr. Mark Winston, a biologist who has worked with insects for more than twenty-five years, writes in *The Biology of the Honey Bee*: "Mounting and copulation are rapid and spectacular, with the drones literally exploding their semen into the genital orifice of the queen. . . . The explosive and sometimes audible ejaculation ruptures the everted [extended] endophallus and propels the semen through the queen's sting chamber and into her oviduct."

When the male inserts his endophallus into the queen, he then flips backward, becoming paralyzed in the process. The force of ejaculation separates the male from the queen and he falls to the ground, where he will die within hours, usually from dehydration. But semen isn't all he's left behind. He's left his penis too. And the next male in line to reproduce with the queen in flight will grab hold of it and throw it away, making way for his own penis and eventual demise.

Some researchers believe the honeybee's detachable penis helps to prevent backflow of semen. But that's probably not its only function. It may also serve as a "mating sign," evidence to workers back in the hive that the queen has in fact mated.

And what happens to the last endophallus? The queen brings it home where it serves to indicate a successful trip. But then, according to Dr. Winston: "A queen returning from a successful mating flight generally is carrying the mating sign of the last drone to mate her, and the workers which greet her lick the sign with their tongues and eventually remove it with their mandibles."

A successful queen returns to her hive with up to 6 million sperm, and she can store them for up to four or five years as she produces hundreds of thousands of offspring to keep her hive buzzing busily along. Now, human males don't leave their penises behind, of course, and the act of intercourse isn't a death sentence, but, like male honeybees, some men expend much energy (think dinner and flowers) to leave millions of male germ cells (sperm) in the hope that just one of them will successfully combine with a single female germ cell (egg). When it comes to the relative value of male and female cells in some reproductive economies, males are plentiful and expendable, but females are rare and precious.

HUMAN PENISES CAN break, but unlike bee penises, they don't break off. There is, however, one small piece of the penis that millions of men—around one-sixth of the world's population—have had removed, their foreskin. Circumcision is the name for the surgical procedure that cuts it away, usually soon after birth. In most cases, the foreskin is removed for religious or cultural reasons; for Jews and Muslims, especially, it's considered a sacred obligation. It's thought that the ancient Egyptians also practiced circumcision. But in a few Western countries, notably the United States, circumcision was routinely performed upon many newborn males in hospitals unless the parents directed otherwise, either because they didn't want their baby circumcised at all or because they were planning on performing a ritual circumcision in a religious ceremony.

Today, there is more and more controversy over circumcision—especially in light of growing understanding about female circumcision, or female genital mutilation, which in some societies involves much more than the removal of an equivalent piece of skin.

Essentially, the foreskin is a thin layer of specialized tissue that covers and protects the head of the penis; it retracts when the penis is erect, or it can be pulled back manually when the penis is flaccid. Like vulvas, penises, and breasts, as well as eyes, noses, fingers, and just about every other anatomical feature, foreskins vary greatly, especially in color and size.

In some men, the tissue covers part of the glans, sort of like a turtleneck pulled up to the ears. In others, it hides the head of the penis completely, hanging over the tip just as the hood of an oversized sweatshirt might droop over the face.

The underside (interior when unerect) of the foreskin secretes a moist substance, part of which can form the cottage cheese–like substance known as *smegma* which makes it easy for the foreskin to glide over the head of the penis. This hot, wet environment can be an ideal place for bacteria and other microbes to creep in and make it their home, unless men are careful to keep the area clean. In some environments, such as the desert, keeping things clean is more challenging. It's easy to imagine grains of sand slipping into the tight space between the foreskin and glans, possibly making for an irritating environment.

Which is why some researchers believe the institutionalized traditions of circumcision grew up in the original desert homes of Judaism and Islam. "Circumcision has a long history in ancient societies of the Middle East," writes Dr. John Hutson, "and is likely to have arisen as an early public health measure for preventing recurrent balanitis [inflammation of the head of the penis], caused by sand accumulating under the foreskin."

If the foreskin is such a breeding ground for infection—to the point where it impedes reproduction—why do we have them in the first place? Anything that can get in the way and completely prevent reproduction should be ejected from the gene pool unless the benefits it confers outweigh its risks.

There are a range of theories about the utility and benefit of foreskins. Perhaps the moisture it secretes facilitates intercourse—and thus, reproduction—by making penetration easier. Another theory suggests that the foreskin acted like a protective cover for the penises of our very early, very naked, ancestors as they roamed through the bush searching for food or shelter. Yet another theory holds that, like the clitoris, the foreskin has a more pleasurable purpose—to make sex feel good, lubricating men's partners, encouraging the intercourse that, until modern science has enabled fertilization in the doctor's office, has been a prerequisite for reproduction.

In a 1999 study, 139 women filled out questionnaires about their sexual satisfaction. In that study, discomfort during intercourse was eleven times more likely with a circumcised partner than an uncircumcised partner. The report's authors hypothesize that this is because, during intercourse, an uncircumcised penis "does not slide, but rather glides on its own 'bedding' of movable skin . . . with minimal friction or loss of secretions." This provides some support for the foreskin pleasure principle, although it's not proof positive by any means.

In Victorian England, circumcision was popularized by its advocates who argued that not only did it help to keep the shameful allure of sexual desire in check, it somehow curbed the sin of masturbation itself, or so they believed.

Cultural unease with sexuality is a driving force behind many of these traditions. And nowhere is that more the case than the appalling practices of female circumcision, a form of genital mutilation, which is still all too present today. According to UNICEF, around 3 million girls are subjected to this brutal practice every year.

There are three basic types of female genital mutilation, and they are performed without any type of anesthesia in almost all cases. In the least extreme case, least being quite relative, the clitoral hood (which is somewhat analogous to the male foreskin) and, sometimes, part of the clitoris is removed. In a more radical procedure, the entire clitoris is removed, along with the inner labia. And in the most extreme form of female circumcision, the entire clitoris and the inner labia are removed and the opening of the vagina is sewn closed or narrowed considerably to prevent intercourse until marriage.

These practices result in a host of dangerous complications. Women who have been subjected to the most extreme type of genital mutilation are 70 percent more likely to hemorrhage after giving birth. These same women are also as much as 55 percent more likely to have a stillborn baby, a result, it is thought, of the increased risk of infection associated with genital mutilation. And, of course, women whose clitorises are excised lose much of their ability to enjoy sexual pleasure and fulfillment.

But how can the cultural tradition of female circumcision be condemned if the cultural tradition of male circumcision is somewhat tolerated, even accepted? There are three critical differences between male and female circumcision. First, while there are potential complications from the procedure of circumcision, as there are with *all* surgical procedures, they are very rare. (This is not to say they aren't serious: infections and penile amputations can and do occur.) Second, while male circumcision is generally performed on an infant in a sterile environment, female circumcision often happens at a later age and in unsanitary conditions. Third, unlike most forms of female circumcision, male circumcision leaves the great majority of pleasure-producing tissue intact.

Or does it? It's hard to know. Ask most circumcised men if their sex life has been hampered by their circumcision and they're likely to look at you like they have no idea what you're talking about. Researchers from Johns Hopkins University recently completed a study of 4,456 Ugandan men; half were circumcised as part of the study, and the other half were left uncircumcised. The study showed virtually no distinction between the two groups in terms of pleasure and satisfaction after two years. 98.4 percent of the circumcised men said they were sexually satisfied, as did 99.9 percent of the uncircumcised men. And 99.4 percent of the circumcised men reported no pain during intercourse, compared to a statistically indistinguishable 98.8 percent of the uncircumcised men.

Of course, the jury's still out. The only people who can truly compare circumcised sex to uncircumcised sex are the very few men outside of scientific studies who have been circumcised after entering adult sexuality, either for religious reasons or for personal preference. Erik Janssen of the Kinsey Institute in Bloomington, Indiana, wants more work to be done: "I think we need quite a bit more data on the direct effects of circumcision on penile sensation," says Janssen. "Is it leading to additional types of stimulation that are more pleasurable? I don't know of really good research on this topic; if there was funding for it, I would study it."

And all the research in the world doesn't mean that there aren't people out there who would rather have things the other way around; there are. There's actually a bit of a cottage industry developing that specializes in reversing circumcision. So maybe there's more to having a foreskin than we currently and fully understand.

But there's another possible reason for the practice of male circumcision. Those ancient desert dwellers may have believed circumcision could prevent infection. If they did, there's new evidence that they were right, although not in the way they imagined, about a disease they likely would not have encountered, and in a way they couldn't have conceived.

For some years Dr. Daniel T. Halperin, of the Harvard School of Public Health, has been pushing the idea that circumcision can help prevent the spread of HIV. And he isn't the only one. More than twenty years ago, not long after AIDS was first identified, a urologist named Dr. Aaran Fink suggested the same thing. For a long time, they and others who shared their belief were ignored or even ridiculed.

But there's nothing like a little data to change people's minds.

A seven-year study in India, conducted from 1993 to 2000, found that circumcised men were seven times less likely to contract HIV. But that study had serious methodological flaws, because it didn't take into account factors like education and economic status, both of which are already associated with a higher incidence of circumcision and safer-sex practices.

Then came a pair of studies in Kenya and Uganda; the Kenyan study found a 53 percent lower incidence of HIV infection among circumcised men and the Ugandan study, 48 percent. The findings were so dramatic, that the National Institutes of Health, which conducted the studies, decided that ethics compelled them to halt the studies midway and offer the uncircumcised participants the opportunity to be circumcised.

More research is still needed, especially because these studies were never actually completed. Modern medical research is generally conducted by studying people receiving a treatment or procedure and comparing them to similar people who do not receive that treatment or procedure; the untreated group is called a control group. If the treatment shows significant positive results early in the study, ethics requires that researchers stop the study and offer the treatment to the untreated group as well. Of course, that also means that you can't be sure what the studies actually would have shown if they had been carried to completion. Perhaps the apparent benefits of circumcision wouldn't have lasted over the long run. Or worse, maybe circumcision would have been shown to increase the risk of contracting HIV. There's no way to know without completing the study.

Now, even if circumcision helps to dramatically reduce the risk of HIV infection, it's by no means a grant of immunity. Circumcised men can still get HIV, and they do all the time. But, combined with safer sex, there's mounting evidence that it really may reduce the risk. Dr. Halperin overlays the prevalence of circumcision and the average number of sexual partners, in a series of African nations, in a striking analysis that bears that out. As he explains, an estimated 15 percent of men in Botswana are circumcised and 65 percent of men report sex with multiple partners; the AIDS rate there is a staggering 25 percent. But in Tanzania, although a similarly high 52 percent of men report high-risk sex, 70 percent of them are circumcised, and the AIDS rate is much lower, at 7 percent. And in Ethiopia, where 75 percent of men are circumcised and only 21 percent report sex with multiple partners, the AIDS rate is 2 percent.

Yet there has been some conflicting research. A recently published study in the *African Journal of AIDS Research* found that in some countries, including Cameroon, Lesotho, and Malawi, circumcision in fact appeared to *increase* the transmission of HIV. And certainly the act of circumcising adults increases the risk of contracting a sexually transmitted infection if the newly circumcised man engages in intercourse before his penis is completely healed.

HIV is a retrovirus that can insert itself into our cells' DNA; this is, in part, what makes it so hard to cure once it has infected someone. HIV is particularly insidious because it targets the cells of our immune system (such as T-cells, which usually protect us from infection) and hijacks them to facilitate its own reproduction. That hijacking eventually destroys those cells, leading to the immune system breakdown we know as acquired immune deficiency syndrome, AIDS. Langerhans' cells, which can also be infected by HIV, are a specialized type of immune cell found in the skin. And guess what part of the body is full of them? The foreskin. So by removing the foreskin you might be removing a prominent site for HIV infiltration. Or so the thinking goes.

HIV isn't the only virus that may be impacted by circumcision. A study published in 2008 in the *Journal of Infectious Disease* looking at 351 men found that uncircumcised men are at increased risk for infection with the human papillomavirus (HPV), including a type of HPV that is involved in cancer of the penis and cervix. HPV is thought to be one of the most common sexually transmitted infections. The multiethnic Hawaiian study found that 46 percent of uncircumcised men had HPV in the glans or corona, whereas only 29 percent of circumcised men were infected. In an accompanying editorial commenting on the study, Dr. Peter V. Chin-Hong of the University of California, San Francisco wrote: "Evidence that male circumcision is associated with decreased penile HPV infection is rapidly accumulating." There are some design limitations to the study, however. It's small, and the circumcised men tended to be older and were more likely to be of Asian descent. When it comes to susceptibility and resistance to disease, genetic differences between ethnic groups can play a large role. For example, people from the eastern Mediterranean and parts of Africa are thought to have developed, through natural selection, a greater resistance to malaria than people who belong to other ethnic groups. So it's possible that the Asians in the study may have had some genetic resistance to HPV infection, and circumcision didn't have anything to do with it.

Another well-known fact may reveal another piece of the puzzle. Scientists have noted for years that women who had uncircumcised male sex partners were at greater risk for cervical cancer. The Hawaiian study also found that even circumcised men with high-risk behaviors for HPV infection (defined as having six or more lifetime sexual partners or sex with prostitutes) were less likely to infect their female sexual partners with HPV. This helps protect their partners from developing cervical cancer, as compared to uncircumcised men. Because of this, some people believe that Gardasil, the quadrivalent vaccine that targets four types (or strains) of HPV may also lower the rate at which men get HPV. These are types 6 and 11, associated with genital warts, and types 16 and 18, which can spawn abnormal cell growth, such as cervical cancer.

If fewer women get infected, there are fewer people who can pass it on. This begs the question: Why don't boys and men just get vaccinated for HPV? Merck, the maker of Gardasil, is hoping to do just that by getting approval to vaccinate boys and young men. Of course, we still don't know if there are any long-term adverse effects from Gardasil. This is always a risk when you introduce a new vaccine or drug. But in the case of HPV, the benefits of vaccination are significant, so let's hope there aren't any serious unforeseen consequences for either sex.

If future studies continue to show that circumcision really does act as a kind of firewall against HIV and HPV, then in countries where there is a very high risk of these infections, perhaps adult men, rather than their parents when they were newborns, should consider circumcision. As with all surgical procedures, circumcision can have serious complications and bad outcomes—and rarely, even fatal ones. In most cases vaccination offers a considerable degree of protection. Cutting off a foreskin is not like vaccination, there's no guarantee of immunity from infections, and, further, it may give some men and their partners a false sense of security. For now, I agree with the American Academy of Pediatrics, which states that routine circumcision is not medically justified.

FOR SOME BOYS in their early teenage years, the first sign that they have entered puberty is a swelling of their testicles. The testicles, which hang in the scrotum—essentially a sack of skin that isolates them from the rest of the body—change very little in size from the time a boy is a year old until the hormone surges of puberty kick off their next and final stage of growth. Adult testicles range in size from a small egg yolk to a large plum. Like the ovaries, the testicles do double duty as parts of both the reproductive system and the endocrine system (the system of glands responsible for production and release and hormones). They produce and store sperm, and from about the sixth week of fetal development they also pump out testosterone, the main sex hormone in males.

Each testicle is partly surrounded by the *epididymis*, a tightly coiled tube that leads from the testicle to the *vas deferens*, which is the superhighway of the sperm release program. The epididymides are entirely inside the scrotum. After the sperm are produced, they move into the epididymis, where they are stored until they leave the man's body through ejaculation.

It's the testicles' role as a sperm factory that explains what at first seems to be one of the most curious features of human anatomy. Why are the keys to our gene pool (and our future generations) literally left hanging outside the body, exposed and vulnerable to injury?

The dominant theory is straightforward—sperm need to be cool.

The scrotum keeps the testicles just a little bit cooler than the rest of the body, in order to create the optimum environment for sperm to flourish and develop normally, a few degrees colder than the rest of the body. That theory is given some real-world support. There is evidence that high temperature, or *scrotal hyperthermia* is damaging to sperm. One of the first things a fertility specialist might tell men with low sperm counts is to stay away from hot tubs.

A recent study by researchers at the University of California, San Francisco, followed eleven men with fertility problems who stopped taking hot baths and getting in the Jacuzzi. Five of the eleven men, nearly half of the study group, saw their sperm counts soar almost 500 percent. And five of the six men who did not benefit from cutting out hot baths were longtime smokers, which is also known to have a negative impact on male fertility. When it comes to male fertility, the stereotypical image of the virile lady's man puffing on a cigar in the hot tub couldn't be further from the truth.

Hot baths and hot tubs "can be comfortably added to that list of lifestyle recommendations and 'things to avoid' as men attempt to conceive," says Dr. Paul Turek, director of the

UCSF Male Reproductive Health Center. Others on the list: smoking, drinking too much alcohol, marijuana, and tight underwear that holds the scrotum close to the body, giving developing sperm the "hug of death" by raising their temperature to the level of the rest of the body.

And a new study by German scientists at the University of Giessen suggests that heated car seats may deserve a place on the list too. The German team outfitted thirty men with temperature sensors on their scrotums and then sat them down on heated car seats, where they remained comfortably toasty for an hour and a half. Sure enough, the average scrotum temperature of the men who sat on heated seats was about a degree higher than the average scrotum temperature of men who sat on unheated car seats for the same period of time. Which may be just enough extra heat to toast sperm as well.

The male body actually has a very specialized climate control system for the testicles. This critical job is actually given to a very small muscle called the *cremaster*. When the temperature drops, the cremaster simply contracts and pulls the testicles up a little closer to the body, warming them; when they need to cool, it relaxes and the testicles drop away from the body. The cremaster muscle isn't only activated in response to the testicles' temperature needs. When a man is stressed, it tightens up and pulls the testicles in snug toward the body, protecting the testicles from possible physical harm.

You can actually test the "cremaster reflex." While you're standing, you or your partner should gently stroke the skin of the inner thigh on one leg. If you have a strong reflex, then the cremaster muscle will contract and pull up the testicle on that side of the body.

Like the cremaster, the *pampiniform plexus* also works to help keep the testicles cool. The pampiniform plexus is a specialized network of veins that brings venous (deoxygenated) blood back from the testicles. The pampiniform plexus acts like a countercurrent heat exchanger (in the same way your fridge or air conditioner at home works): it takes heat away from the blood that is headed to the testicles via the testicular artery. This means that the blood that supplies the testicles with oxygen reaches them at a lower temperature.

During development, testicles normally start out inside the abdomen and descend before birth. But they don't always descend. As many as 30 percent of boys born prematurely and 3 to 4 percent of boys born full-term have at least one undescended testicle. So, if your child has an undescended testicle, it's important to talk to your doctor about it: if left untreated an undescended testicle can permanently lose the ability to make sperm and is more likely to become cancerous. In about 65 percent of newborn boys with the condition, the testicles descend naturally by about nine months of age. If they don't descend on their own, the condition can usually be corrected with a surgical procedure called *orchiopexy* or with a hormone injection.

When it comes to sexuality, a common complaint *some* men have regarding their testicles is the uncomfortable condition known as pelvic congestion or, more colloquially, "blue balls." When a man is sexually aroused, there is an increase of nitric oxide in the penis, which leads to an erection. Nitric oxide is a vasodilator, which means that it causes the smooth muscle that surrounds arteries to relax, allowing the penis to flood with blood. At the same time, the veins leading away become constricted, which is what allows the increased blood flow to fill up the penis and expand it. But it's not just the penis itself that's affected by this process, which is called vaso-congestion. The entire genital area is flooded with blood, the drainage pipes are blocked, so to speak, and everything swells. A quick way to counter this change is an orgasm: the system then shifts into reverse, the arteries constrict, the veins dilate, the flood of blood recedes, and everything returns to normal.

But, if there is no orgasm, there is no immediate signal, no narrowing arteries, and no widening veins—just a whole lot of blood sitting still in the genital region. The actual mechanics behind the testicular pain or ache that some men report after prolonged excitement without orgasm is not completely understood. One idea is simply that they ache because all the extra blood places increased pressure on the highly pain-sensitive testicles. Another related idea is that the tissue in and around those vessels becomes starved for oxygen, which is certainly known to cause pain. That's what we believe causes angina pectoris, the chest pain caused by a lack of oxygen in the heart muscle. The word *angina* actually comes from the Greek "to strangle." Whatever the cause, this much is sure—the phenomenon of blue balls is real, they really can ache—and men are not the only ones to experience this; some women actually complain of a similar effect localized to the lower pelvic region. Eventually, of course, the body realizes that there won't be an orgasm and signals the surrounding blood vessels to begin the drainage process to restore normal blood flow. In reality, the only real blue balls found in nature belong to the African vervet monkey, which has a bright blue scrotum that looks like nothing so much as a pair of big robin's eggs.

It has often been suggested that the size of one's testicles has some bearing on courage or manliness. But, despite the cross-cultural popularity of this belief, size may be more connected to a species' appetite for promiscuity.

It's long been known that the more promiscuous females of a given species tend to be, the larger the males' testicles relative to body size. For example, female gorillas are somewhat monogamous, and male gorillas have especially small testicles relative to their body size. Chimpanzees, on the other hand, are seriously promiscuous—and their testicles are ten times the size of those of gorillas on a relative basis. And humans? We're comfortably—or uncomfortably?—in between.

Why the difference in testicle size? It's all about the competition. Sperm competition, that is.

If a female mates with multiple males, the odds that any one of her partners is going to be the one who successfully gets her pregnant drop significantly because the sperm from all those other males are competing to fertilize her egg. How does the male increase his odds of passing his genes on? One way is to overwhelm the competition with numbers. Simply put, the more sperm the male can successfully ejaculate into the female, the better the chances of that male's sperm finding their mark. And bigger testicles produce and hold more sperm.

Yet having large testicles can come at a cost—especially if you're a bat. Scientists who studied the correlation between anatomical size and behavior in more than three hundred species of bats came to the following conclusion: "Because relatively large brains are metabolically costly to develop and maintain, changes in brain size may be accompanied by compensatory changes in other expensive tissues."

The researchers found that in the species of bats in which the females are wanton, males were shortchanged when it came to brain size. Don't feel sorry for the males though. They were given a big boost when it comes to reproduction. In this case, the expensive tissues they're talking about happen to be testicles.

Scientists have found the correlation between testicle size and mating system across the primate family, and in some other animal groups, such as birds. The more likely females are to have multiple mates when fertile, the larger male testicles are likely to be relative to body size.

But having larger testicles may not be the only way to ensure reproductive success. In 2008, scientists from the University of California at San Diego and Irvine discovered that when it comes to speed, the sperm of chimpanzees and rhesus macaques (primates that tend to be on the more promiscuous end of the spectrum) are markedly faster swimmers than human and gorilla sperm. According to Jaclyn Nascimento, one of the researchers involved in the study, "Rapidly swimming sperm cells would be evolutionarily favored when the mating pattern is polygamous, and that is consistent with our measurements of chimp and rhesus macaque sperm."

Besides overwhelming the competition with both sheer numbers and speed, there's another weapon in the biological arsenal used to fight promiscuity. Some primates have evolved stickier semen. Having sticky semen can work as a physiological stopper; in fact, chimpanzee semen is so thick and firm that it can actually form a plug. This is thought to come in handy when trying to block sperm from subsequent ejaculates of rival chimps. A gene found in primates called SEMG2 encodes for a protein called semenogelin II, which, like corn starch added to a thin soup, is thought to result in thicker semen. The SEMG2 gene from chimpanzees is thought to have experienced a rapid degree of evolution to keep up with the sexual competition (in this case promiscuity) that they experience on a day-to-day basis. "It's similar to the pressures of a competitive marketplace," says Dr. Bruce T. Lahn of the University of Chicago. "In such a marketplace, competitors have to constantly change their products to make them better, to give them an edge over their rivals—whereas, in a monopoly, there's no incentive to change."

So how does human semen compare? The evolution of the human SEMG2 gene has been found to be midway between chimpanzees and gorillas, just like testicular size. In fact, research published in 2008 found that Eastern Lowland gorillas (*Gorilla beringei graueri*) are so confident when it comes to paternity that their SEMG2 has become a *pseudogene,* a piece of defunct genetic currency that the cells can no longer cash in to form a usable (functional) protein.

It's important to note, that when faced with competition from other males, not all species have simply, over time, increased the number of sperm they ejaculate. Scientists have observed that in some species, such as the European bitterling fish (*Rhodeus sericeus*), the number of sperm are actively reduced when there is too much competition. The reason for this is conservation: if the chances of success go down (because there are too many males), then it may be a waste of resources to spend too much energy on sperm. But who knows—some male fish do get lucky, so it may still be worth a shot, even if it's with less sperm.

NOT ALL BOYS experience nocturnal emissions or "wet dreams." For those who do, they can actually be a positive sign that everything is coming together, from semen production to the complex wiring of the nervous system that allows for arousal and culminates in ejaculation and orgasm. Like a girl reaching menarche, the first time a boy ejaculates, he's far from fully fertile; the concentration of sperm in his ejaculate is still very sparse, and it will be anywhere from one to three years before he reaches full fertility. The average ejaculate of a mature male contains between 40 and 150 million sperm, while a young man's first ejaculate may contain significantly less—but he's getting there. None of which means he's incapable of impregnating someone, of course.

Sperm are little marvels. Impossibly tiny—they're among the smallest cells in the human body—they are also fast, efficient, and adaptable. Sperm are just one five-thousandth of an inch long; egg cells (ova), on the other hand, are among the largest human cells, about thirty times larger than the heads of sperm.

Sperm make up around 1 percent of ejaculatory fluid, but when it comes to semen it's really all about them. The followers of the Greek thinker Pythagoras believed that semen was a "clot of brain containing hot vapour," but in fact 99 percent of it is composed of sugars, fats, proteins, and alkaline fluids that variously serve to provide energy, security, assistance, or safe passage to sperm as they set off on their journey to penetrate an egg.

Here's what happens when a man ejaculates: sperm move through the vas deferens, which carries them up into the body toward the prostate gland. Right before they reach the prostate, the vas deferens merge with ducts from a pair of glands called the *seminal vesicles* to become the ejaculatory ducts. As the ducts merge, the sperm are combined with fluid from the seminal vesicles that includes amino acids, vitamin C, sugars to provide energy for the sperm, and prostaglandins, compounds designed to mildly suppress the female immune system to prevent it from attacking the sperm. The fluid from the seminal vesicles makes up about 60 to 70 percent of semen.

The ejaculatory ducts then pass through the prostate gland, which secretes additional fluid that is highly alkaline, which will help to neutralize the natural acidity of the vaginal canal, ensuring better conditions for sperm to survive as they enter the urethra. This prostatic fluid is also rich in zinc, and makes up about 25 to 30 percent of semen. From there the combined sperm and seminal fluid pass by the opening to the bulbourethral glands. These glands have already done most of their work, emitting some of the clear liquid known as pre-ejaculate that clears the way for semen by cleaning up any traces of urine or acid in the urethra. Finally, the sperm is pleasurably shot out of the penis, searching to fertilize an egg.

For most sperm, of course, that's an impossible dream. The numbers alone make that clear—with 150 million related competitors and one (maybe two) eggs *if*—and it's a big if—your partner is ovulating, the odds are pretty stiff. On top of that, most sperm just aren't up for the job. Dr. Harry Fisch, a urologist at Columbia University Medical Center and a specialist in male fertility, states that "only a perfectly normal sperm can penetrate an egg and the majority of sperm are abnormally shaped." "Abnormal" can mean two heads, no tails, or just no ability to move at all. According to Dr. Fisch, a man with 15 percent viable sperm is doing very well indeed.

But every once in a while some lucky sperm swimming upstream detect an egg meandering downstream through the Fallopian tubes. When they sense the egg's chemical signature, they switch to what Australian sperm expert Dr. Moira O'Bryan calls a "crazed figure-eight motion." A few of those actually get close enough to the egg to penetrate its outer shell and, sometimes, almost magically, one of them does.

Just often enough for all of us to be here.

BY THE WAY, there's one interesting way for a man to tilt the odds toward reproductive success: pornography. But not just pornography involving women—it has to involve the competition, so to speak. Think back to our discussion of sperm competition relative to testicle size. Larger testicles mean more sperm, and this increases the odds of reproductive success.

But having more sperm isn't the only way to improve your odds in the fertilization race—faster, stronger sperm can make a real difference, too. A 2005 Australian study showed that when men looked at pornographic images of two men and a woman together, they produced significantly better quality sperm than when they looked at images of just women. Evolutionary biologist Leigh Simmons, one of the researchers of the study, stated that "males ejaculate more sperm, or sperm of better quality, when the risk of sperm competition is high. . . . We found men viewing images containing both men and women had higher sperm motility in masturbatory ejaculate compared to men who were viewing images of just women alone."

Although there is still *a lot* of confirmatory work that needs to be done, this research is really exciting because it suggests the very strange and hard to believe possibility that men can actually dictate the quality of sperm they ejaculate.

BY THE WAY, just because most of semen is a sophisticated support team for sperm success, it often has more of a starring role sexually, especially where oral sex is concerned. As far as semen is concerned, you are what you eat. Yes, it's true: what you eat affects how your semen tastes. Foods and beverages with bitter flavors, like coffee and alcohol, can make your semen taste bitter. Foods with more delicate flavors, like pineapple, celery, and melon, can make your semen taste less "strong." And sure enough, someone with a diet rich in meat is likely to produce semen that, according to some, is thicker and gummier. For the lightest semen of all, vegetarians are the connoisseur's choice.

But don't take my word for it. The BBC actually commissioned a taste test to check out the theory in the real world. Three couples participated. The men were put on one of three specific diets for three days and their partners were not told which one. One man ate all seafood, another ate hot and spicy food, and the third was put on fruits and vegetables. The women were asked to identify their partner's diet after tasting a sample of his semen from a plastic test tube on camera. Sure enough, the woman whose partner ate all seafood identified a fishy quality, and the woman whose partner ate all fruits and vegetables found his semen to be positively "lighter" than it had been. Incidentally, the woman whose partner was on the seafood diet demanded an immediate return to her partner's cheeseburger-eating ways. She hates fish.

By the way, we're not the only animal that engages in oral sex. It seems that some types of macaques, cheetah, hyenas, gibbons, and even goats, perform oral sex and even swallow semen.

Semen can actually do far worse than leave a bad taste in your mouth. Some people are actually allergic to their partner's semen, and when they are exposed to it through any type of sexual intimacy, it can cause itching, burning, and, in rare cases, difficulty breathing. A semen allergy can be a response to specific proteins in your partner's semen alone or the result of a more general allergy to all semen. The good news is that semen allergies can be treated, and you can usually be desensitized. How? By repeated exposure to the allergy-prompting semen. But take note: a serious semen allergy, like any other allergy, can be dangerous, even life-threatening, and desensitization should only be done in consultation with a doctor.

And there's a little more good news. Once desensitized, you need regular maintenance procedures to preserve the reduced sensitivity. Otherwise known as regular sex.

i'm so excited and i just can't hide it

Millions of words—in books and on blogs, in magazine articles and advice columns—have been written about mastering the search for Ms. or Mr. Right. And millions of people have wondered why they keep dating the wrong man or woman when they think they know what's really good for them. They think they know what they need; the problem is, what they *need* isn't always what they *want*.

So why do we want what we want anyway? How much of what turns us on is hardwired?

Like everything else, attraction and arousal (and possibly love, for that matter) are the products of millions of years of finely tuned biological engineering. And there's really only one goal behind the engineering—to get you to have sex.

As you'll see, much of what we are preprogrammed to find attractive may be connected to what it tells us about the health, fertility, and compatibility of potential mates. Genetic compatibility, that is. Nature isn't really concerned about similar political views or favorite movies, although it does place stock in appearances. From the standpoint of survival, in some sense it really cares that we have strong offspring, and that they get what they need to grow up and give us grandkids in turn. But genetic compatibility only gets us halfway there in terms of successful offspring—it can give us healthy babies, but those babies need parents to protect and nurture them into maturity. And that's where love comes in. Falling and staying in love—the pair bonding that keeps a couple together long enough to have, raise, and care for children—almost certainly involves chemical processes that are a product of millions of years of evolution.

A new study published in March 2008 in the journal *Evolution and Human Behaviour* shows that being in love with someone actually works to dampen the sexual appeal of people of the opposite sex that we might otherwise find attractive. In another study, Florida State University researcher Jon Maner took two groups of heterosexual college students who "were currently in a committed romantic relationship" and showed them rapid-fire bursts of images depicting very attractive and average-looking men and women. Before watching the images, each group wrote essays. The first group of students watched the images after writing essays about extreme happiness. The second group wrote essays about moments when they felt extreme love toward their partners. In the group that was primed by writing about happiness, the participants in the study seemed to pay about the same amount of attention to attractive people and average people. But in the group that wrote essays about moments of extreme love toward their partners, their attention, in the words of the researchers, "was captured substantially less by attractive alternatives than by other targets." The researchers' theory is that concentrating on the love you have for your partner may block the normal reflexes that might otherwise cause you to consider other potential "attractive" partners. This may be a mechanism that evolved to keep couples together. If you keep seeking alternatives to your partner, the likelihood of building a lasting and successful relationship, especially one that leaves you with children, diminishes. There will always seem to be a more attractive prospect than the person you love, so love may work to stop you from looking. Maner further described his findings:

> We found that when people just thought about being in love with their current partner, their visual attention got repelled, rather than grabbed, by an attractive member of the opposite sex. [That] happens at the very initial stages of visual processing, at the very first moment they are aware of the photo.

Love was actually working to limit individual receptivity to potential sexual partners that posed a threat to existing mates. Joseph Forgas, a psychologist from Australia's University of New South Wales, explained the potential significance of this research:

> Psychologists have long had a problem explaining the functions of romantic love, a very strong emotion that sometimes

seems to take over our lives and lead to what appear to be irrational feelings and actions. What these studies suggest is that romantic love serves a very important function, tempering our natural desire to pay attention to, and to continuously seek out, the best available mate.

BEFORE WE PLUNGE into exactly what turns us on and why, let's pause for one further twist. A lot of what we find sexually attractive may be hardwired, but I need to mention a little wrinkle. What we find attractive can change. For some people, who they find sexually attractive can actually be different depending on the time of the month.

Numerous studies have shown that women's preferences shift with their menstrual cycles. When they're at peak fertility, the few days before, during, and after ovulation, they lean heavily toward supermasculine types: think tall, deep-voiced, darker-skinned, and muscular. Many of these traits can act as a sort of well-groomed genetic résumé, telegraphing the health and fitness of a potential partner and suggesting his suitability as a source for future children.

"Women know they have attractions that come and go, but they probably don't realize that these urges are tied to their cycle—as well as our evolutionary past," writes Martie Haselton, a UCLA researcher who has studied the connection between attraction and fertility. "They just know that suddenly one day they're attracted to their hunky neighbor or handsome co-worker. . . . Ancestral women who were attracted to these features produced offspring who were more successful in attracting mates and producing progeny. The legacy of the past is desire in the present." In other words, at ovulation time, good traits mean good mate.

Of course, good traits aren't all that high levels of testosterone can mean. Higher levels of testosterone correlate with higher levels of aggression, a trend toward dominance, and a lack of fidelity. Which is possibly why when women are *not* ovulating, they tend to be attracted to a different set of characteristics, like softer features and larger eyes, which some link with stability, nurturing, and other qualities that suggest someone would be a good partner and parent.

A recent study led by Dr. Haselton seems to indicate that women actively work to make themselves look more attractive when their fertility is highest. Researchers recruited a group of adult women of childbearing age, between eighteen and thirty-seven, and photographed them twice: first, when they were close to ovulating, when fertility is highest, and second, when they were close to menstruating, when fertility is lowest. Dr. Haselton and her team then showed the photos to a group of volunteers, asking them, "In which photo is the person trying to look more attractive?" The researchers wanted to understand if the women actually changed the style of their clothes and accessories in a way that related to their fertility levels. The women's faces were blocked out so variations in facial expressions wouldn't distract the volunteers from the women's clothes, jewelry, and so forth. Sure enough, overall, the volunteers thought that the women were trying to look more attractive in the photos that were taken around the time the women were ovulating. It's worth noting that this fertility effect appeared even though all the women photographed in the study described themselves as being in committed relationships with men.

It's almost as if women are looking *to* mate when they're ovulating, but *for* a mate when they're not. And by the way, when women are looking for the right set of traits in a potential father, they're willing to look a little farther and a little wider than usual. Two recent studies have shown that women are more likely to cheat when they're ovulating.

"We found that women were most attracted to men other than their primary partner when they were in the high fertility phase of the menstrual cycle," Haselton has said. "That's the day of ovulation and several days beforehand."

A related UCLA study published in *Evolution and Human Behavior* found that women were more likely to fantasize about men other than their partners when their fertility was highest. Both studies, however, found an exception—women with highly attractive partners did not experience the heightened desire to stray or fantasize about it.

By the way, you undoubtedly recognize that the effects of cheating can be quite complicated, but if you think you've heard all the ways cheating can change things, think about this: cheating can lead to twins with different dads. That's right. It's called *heteropaternal superfecundation*, and here's how it works. As in the normal conception of fraternal twins, a woman releases two eggs when she ovulates. But instead of having sex with only one man during her ovulatory period, this future mom of half-sibling twins has sex with two (or more) men and each egg is fertilized by a different man's sperm. So they share the same mother but have different fathers. If you're curious as to how common bipaternal twins may be, one paper suggested that one in four hundred pairs of fraternal twins born to white married women in the United States may actually have different fathers.

Before any woman considers using ovulation as an excuse for her promiscuity or any man starts tracking his partner's menstrual cycle out of paranoia, they would do well to consider the view of Elizabeth Pillsworth, a UCLA assistant professor of journalism and psychology who coauthored one of the studies. "Whether they [these desires] translate into unfaithful behaviors is a matter of their own choosing. Cheating is a choice," observes Pillsworth. "I hope the message women get is that they can use this information to realize their biology is toying with their desires and to ask themselves, 'Am I going to let that run my life, my sexual decision-making?' For the men I would say not to be too fearful of these findings. While women may notice other men during this part of their cycle, unfaithful behavior is relatively rare."

In other words, what you want may not be entirely up to you. But what you do about it is.

WHEN IT COMES to attraction and arousal, there are two senses we obviously rely upon—sight and touch. But anyone who has buried their head in their departed lover's pillow to inhale the scent that remains can tell you, smell plays a major role.

Rachel Herz is a psychologist and the author of *Scent of Desire*. In *Psychology Today*, her colleague, Estelle Campenni, describes telling Dr. Herz a story familiar to many women:

I knew I would marry my husband the minute I smelled him. I've always been into smell, but this was different; he really smelled good to me. His scent made me feel safe and at the same time turned on—and I'm talking about his real body smell, not cologne or soap. I'd never felt like that from a man's smell before. We've been married for eight years now and have three kids, and his smell is always very sexy to me.

How does it work? Much of the research around smell involves a search for human *pheromones*. Pheromones are chemicals that trigger specific behaviors in many organisms. Scientists have established their existence in thousands of species. But whether there are any human pheromones is a matter of some controversy.

"As of now, a lot of the claims that people might be making about human pheromones are simply not true," explained Professor Charles Wysocki, a neuroscientist from the Monell Chemical Senses Center in Philadelphia. He went on to say, "There's no study that has yet led to the isolation of a true human pheromone. But that doesn't mean that they can't exist."

What this really means is that we don't know very much about the subject yet. "It's like what we used to know about willow bark tea." Explained Wysocki, "We knew it could bring down a fever and control mild pain, but we had no idea that it was because it contains salicylic acid, the basic chemical building block of aspirin." That's where we are with smell and human sexuality. We think it does something, but we just don't know what.

Most likely it does a lot.

THE GRANDDADDY OF research into scent and sexual attraction is a biologist at Switzerland's University of Lausanne named Claus Wedekind. In 1995, he gave forty-four men new T-shirts and asked them to sleep in them two nights in a row, ensuring that the shirts were steeped in their sweat and accompanying scent. He also gave them odorless soap and aftershave to ensure that nothing masked the odor of each man's natural "perfume." Wedekind then asked forty-nine women to smell each shirt and rate their attractiveness. Time and again, volunteers were more attracted to the smell of shirts worn by men who had immune systems that were somewhat different from their own (more on how this works in a few pages), especially a group of very important genes that make up a key part of our immune system: human leukocyte antigen system, or HLA.

HLA acts something like a programmer at the heart of our immune system; it codes for the proteins that our immune system uses to recognize and to fight outside invaders and threats. Many of the genetic variations that doctors examine to determine if a recipient will reject or accept an organ or tissue transplant involve HLA genes.

There can be millions of genetic variations in HLA gene combinations. The more relative diversity in your HLA, the more robust or flexible your immune system is, because it's going to recognize and deal with a wider variety of potential pathogens, making you both more resistant and better equipped to overcome infectious disease. Relative diversity, because it's possible that there are some regional specializations to HLA—specific genes to combat diseases that are specific to a given environment.

All of which means that, if you want to give your potential child the best odds for the strongest possible immune system, you need to combine your HLA system with another that's somewhat dissimilar. Unlike many other classes of genes, HLA can be co-expressed (which means that both copies are expressed). Having many different variants of these genes may end up providing just the right combination to deal with microbial threats. Children from parents with a very similar HLA profile, like close relatives, may not provide the best genetic opportunities to fight off microbial infections.

Farmers face a similar problem. When they attempt to increase their yields by planting an entire field with genetically similar seed (called *monoculture*), the resulting plants are more susceptible to disease because they lack the genetic diversity to fight off a wide range of pathogens.

Students of history may recall the Irish potato famine in the mid-nineteenth century. Potatoes were the primary cash crop in Ireland at the time. Most of the plantings were of the lumper variety—a monocrop. This created a situation ripe for disease. When the entire potato crop failed as a result of a fungus plant pathogen called potato blight, or *Phytophthora infestans*, the economy collapsed and thousands of Irish workingmen, women, and children died of starvation. One way some farmers try to avoid making the same mistake today is to plant different varieties of the same plant. In the same way, we avoid having children with very close relatives to ensure a more dynamic reshuffling of our genes.

In fact, having a reproductive partner with very similar genes, called *consanguinity*, may make it even more difficult to conceive, and it is also thought to increase the risk of bearing a child with congenital diseases. The increased difficulty in conception may be an outgrowth of the same phenomenon: if highly similar genes or relatedness between partners increases the risk of birth defects, it may also increase the risk of miscarriages resulting from abnormal embryos and fetuses.

Having children with someone with a completely dissimilar HLA profile, meaning very big differences in their HLA genes, may not be the best idea either. If two people with vastly different HLA profiles have children, they might not pass on an HLA combination that evolved in response to the microbial environment they live in. This means giving up on the best chance to fight off infections and infestations that are specific to a certain geographic location. In reality, of course, it's far more complex, but you get the idea—partners with diverse HLA are giving their potential offspring an immunological leg up, and partners whose HLA genes are too closely related may have difficulty conceiving.

Much of my own research has dealt with the relationship between certain HLA genes and resistance to infections, but, when it comes to olfaction and attraction, I've always wondered what it is exactly that people smell when they are sniffing out a potential partner. When I spoke with Herz, she explained how she believes that HLA manifests itself as body odor:

> Your body odor, whatever it is, is reflective of your particular HLA. However you smell when you come out of the shower, before you put anything on, is a reflection [of HLA] from the point of view that the proteins that your HLA is coding for are being degraded by the bacteria that are on your skin's surface. And the by-product of that degradation is what produces your very own body odor. And because you have different proteins expressed as a function of your HLA uniqueness, then you have a specific, distinctive body odor.

The good news about scent and sexuality is that there are always going to be people who like your particular odor. Because the attractiveness of your smell depends on the genetic makeup of the smeller, and their past exposure, and because there is so much genetic variety, your smell is bound to have many matches out there. "There's no Brad Pitt of smell," says Herz. "Body odor is an external manifestation of the immune system, and the smells we think are attractive come from the people who are most genetically compatible with us."

Not only does HLA compatibility have a bearing on a couple's suitability as potential parents; it also has implications for their suitability as potential partners for life. University of New Mexico researcher Christine Garver-Apgar studied the relationships of heterosexual couples in the context of the relative similarity or dissimilarity of their HLA patterns. The couples provided information about their relationships and sexual habits, and then matched that information against genetic testing that documented their HLA patterns. They found that the more HLA genes a couple shared, the less sexually responsive the women were to their partners and the more they cheated on them. In fact, Garver-Apgar found that there was a direct correlation between the number of HLA genes a couple shared and the odds that the woman would cheat. Fifty percent HLA genes in common? Fifty percent chance the woman was fooling around. What about men, you ask? Most of the studies involving smell concentrate on women, who naturally have a much keener sense of smell than men. Women's preferences also seem to be much more affected by smell.

But relying on sense of smell alone is not foolproof when it comes to mate selection. Earlier this year a British couple sought to have their marriage annulled after discovering they were fraternal twins. They were adopted by different families as babies and grew up without any knowledge of each other. If every nose were a genetic bloodhound sniffing out and rejecting similar genes, then your twin ought to smell like a skunk. Fortunately for family unity, it doesn't really work that way; we tend to be comforted by the familiar smells of family, even if we want something else in a mate.

Though women are more sensitive to male smells in general, men are especially sensitive to women at a specific point in their menstrual cycles. When? When they're ovulating, of course.

Professor Devendra Singh, of the University of Texas at Austin, conducted his own version of the T-shirt test. He gave two new T-shirts to about two dozen women. They slept in one shirt during the most fertile part of their menstrual cycles, on days 13 to 15; they slept in the second shirt on days 21 and 22 when they were no longer actively fertile. As in the Wedekind study, the women volunteers were asked to avoid perfumes, scented soaps and shampoos, and even pungent foods, like garlic. Sure enough, when men were asked to sniff the shirts and pick their preference, they picked the smells from the fertile phases over and over again.

Smell seems to affect the process of attraction in direct relation to nature's expectations of us in the reproductive process. Since the biological cost of reproduction is so much higher for females than for males, it shouldn't be a surprise that women may be wired to sniff out men who will provide the right traits to give them the healthiest babies. And men in turn may be wired to sniff out women who are ready to make babies.

Interestingly, the compounds that seem to exert the strongest sexual pull are gender-specific. That is, male compounds trigger a sexual response in women, and female compounds trigger one in men—with one important exception. And the exception really does seem to prove the rule.

Homosexual men have the same reaction to certain male odors that heterosexual women have, not just subjectively, but in terms of measurable brain activity.

Ivanka Savic-Berglund and a team of researchers at the Karolinska Institute in Stockholm, Sweden, used a brain-imaging technique to examine the responses of a group of straight men, a group of straight women, and a group of gay men, to two odors, one from men and one from women. The male compound was a testosterone-related chemical found in men's sweat, and the female compound was an estrogen-related chemical found in women's urine.

Most smells activate specific areas of the brain that are known to govern how we receive and process scents. And that's exactly what the sweat compound did when heterosexual men smelled it—it activated the smell-related areas of the brain. Same for women and the urine compound. But when straight men were exposed to the female compound, brain imaging revealed that not the smell-related zones but the hypothalamus, which controls sexual behavior, kicked into overdrive. The same thing happened when women were exposed to the male compound. But what was especially noteworthy was the response of gay men. They responded to the male compound exactly as the straight women did. Instead of engaging the smell-related zones of the brain, the brain imaging revealed that the male sweat compound energized the hypothalamus of gay men.

Dr. Savic followed up this study in 2006 with a similar examination of twelve heterosexual women, twelve heterosexual men, and twelve homosexual women. As in the previous study, the heterosexual men processed the male odor in the normal scent regions of the brain and the female scent in the sex-related area of the hypothalamus. The heterosexual women did the opposite. In the case of the homosexual women, the male sweat-derived odor activated the olfactory centers of the brain as expected, but the female urine-derived scent activated both the sexual areas of the hypothalamus and the normal olfactory centers of the brain. The results were not quite as straightforward as with the gay men in the previous study, but the female scent did activate the sexual areas of the brains of homosexual women in a way that it did not in the heterosexual women.

Charles Wysocki, the neuroscientist from Philadelphia's Monell Chemical Senses Center, completed a study that asked "odor evaluators" in four categories—heterosexual men, heterosexual women, homosexual men, and homosexual women—to indicate their preference among odors collected from "odor donors" in the same four categories. None of the odor evaluators were odor donors.

The researchers asked the donors to go through a "washout" phase for nine days in order to cleanse their bodies of foreign odors; they used odorless soaps and shampoos, avoided pungent foods like garlic and curry, and didn't shave their armpits. For the next three days, they wore cotton gauze pads under their arms. The pads were then cut into pieces and combined with pads from other donors in the same category (heterosexual men or women, homosexual men or women) and put into plastic squeeze bottles. This gave the researchers a kind of hybrid odor sample for each class by reducing the impact of any individual odor.

When odor evaluators were asked to choose between different odor samples, a few clear patterns emerged. Here's some of what the report said:

> Heterosexual males, heterosexual females, and lesbians preferred odors from heterosexual males over odors from gay males; gay males preferred odors from other gay males. . . . Heterosexual males, heterosexual females, and lesbians over the age of 25 (but not those ages 18–25) preferred odors from lesbians over odors from gay males. . . . Finally, gay males preferred odors from heterosexual females over those from heterosexual males.

If evaluators were selecting odors randomly, no clear patterns among the gender and orientation groups would have emerged, of course. Clearly, some complex interaction between orientation, attraction, and odor affected the participants' choices. By the way, if you're wondering about the role of HLA, the genes that play a key role in our immune system, this research examined some of the broad strokes of odor and sexuality—specifically as it relates to sexual orientation. The relation of HLA, odor, and attraction seems to involve a finer filter—the genetic compatibility of potential parents, not fundamental gender attraction. But this clearly demonstrates that there is much more to smell and attraction than HLA.

What isn't clear from any of these studies, of course, is whether the preferences participants indicated were a result of their sexual orientation or involved in determining their orientation.

"Our study can't answer questions of cause and effect," says Dr. Savic. "We can't say whether the differences are because of pre-existing differences in their brains, or if past sexual experiences have conditioned their brains to respond differently."

What is clear is this: smell is intimately involved in intimacy. The next time somebody tells you they just don't have the right chemistry with someone, they might mean it literally, even if they don't know it.

WHILE THE SMELLS we find sexy may be as varied as our DNA, sexy sights are another matter entirely. There actually are some pretty universal standards for what humans tend to find visually attractive, and they start with a cliché—tall, dark, and handsome.

Across cultures and continents, many women tend to be attracted to men who are relatively darker than others in their group. And there's a perfectly good reason—they are likely to have healthier sperm.

Folic acid, or folate, is an essential nutrient found in leafy greens such as spinach; it gets its name from the Latin word for leaf, in fact. Folate is critical to the healthy production of new cells. It's especially important during periods of rapid cell growth, as in pregnancy, which is why many doctors advise women to take it as part of a prenatal vitamin package when they're trying to get pregnant (even up to a year in advance), to prevent neural tube defects such as spina bifida.

A recent study published in 2008 in the journal of *Human Reproduction* by a team at the University of California at Berkeley, suggests that folate is important for male reproductive health as well. Men with higher levels of folate had higher percentages of healthy sperm.

In an average healthy man, up to 4 percent of the sperm in his ejaculate has the wrong number of chromosomes, called *aneuploidy*. High percentages of aneuploidy can impact fertility and has been linked to miscarriage, birth defects, and genetic disorders such as Down syndrome. The Berkeley study showed that men who consumed the highest levels of folate through food, vitamin supplements, or a combination of the two, had as much as 30 percent less aneuploidy in their sperm than men with the lowest levels of folate. When you consider that a normal adult male produces sperm at the astronomical rate of around 100 million per day, it makes sense that a vitamin that promotes healthy cellular reproduction would be good to have around.

But where does tall, dark, and handsome come in? Ultraviolet rays destroy folate. Darker skin is achieved by cells called *melanocytes* that produce different types and amounts of melanin, (naturally occurring pigments that absorb ultraviolet radiation and releases it as heat), protecting the body from its harmful effects. So the darker a man is, the more likely he is to be protected from ultraviolet rays, which means the less folate is destroyed, which means the healthier his sperm.

Tall, dark, handsome, and ready to reproduce.

BUT THIS IS not the only cliché that has a bearing on what we find visually attractive. What about the old adage about truth from the mouths of babes? Well, babies have an opinion when it comes to physical beauty.

According to the magazine *New Scientist*, researchers at Britain's University of Exeter presented a series of facial pictures to a group of adults and had them rated for attractiveness on a scale of 1 to 5. They then paired photos that were similar in terms of composition, lighting, and contrast but at opposite ends of the attractiveness scale. The pairs were presented to babies less than a week old. We don't know why, but in almost every case the babies spent considerably more time looking at the face with the higher attractiveness rating.

According to Alan Slater, an associate professor from the University of Exeter who studies development of perceptual and cognitive abilities in infancy, "Attractiveness is not in the eye of the beholder, it's innate to a newborn infant." He theorizes that what makes an attractive person attractive is how closely he or she resembles a kind of idealized, prototypical human face. The more someone looks like this theoretical prototype, the more likely others are to consider him or her as a potential partner. Slater's theory certainly agrees with what researchers have long known about faces—when you blend the features of hundreds of random faces, the resulting "average" face is inevitably beautiful. According to the theory, then, it's not the size of your nose on its own that affects your attractiveness; it's how much the size of your nose deviates from this "average" ideal. So how do babies measure attractiveness? Slater believes babies are hardwired with that prototypical image of the "average" face: "Babies are born with a fairly detailed representation of the average human face that helps them recognize familiar faces and also helps them learn about the social world," says Slater.

If babies can really recognize beauty, we ought to be able to quantify it, don't you think? But as Charles Darwin, the standard-bearer for evolutionary theory, has said, "It is certainly not true that there is in the mind of man any universal standard of beauty with respect to the human body." And some researchers, scholars, and philosophers might agree.

In 2007 researchers from the University of Stirling, Harvard University, and Florida State University published a study that examined standards of beauty on two continents. The results would seem to disagree with Darwin and others, and further confirm what other studies over the last two decades have indicated: there is in fact one universal quality that people find attractive, on the plains of Tanzania, in the streets of London, and around the globe: symmetry.

The research team showed eighty British people and forty members of Tanzania's Hadza tribe, one of the only hunter-gatherer cultures still in existence, the same series of faces. In both groups, there was a clear and marked preference for symmetrical faces. Symmetry, of course, means exactly that—the same on both sides. And it may be that external symmetry is the best visual indication that the person is carrying a combination of genes, which also leads to proper internal development of organs and blood vessels. In other words, developmentally speaking, everything worked out.

The roots of our preference for symmetry lead to the same place as all the other preferences we've discussed. To some extent, symmetry advertises the pedigree of its owner; the right combination of genes and the exposure to a good environment.

In nature, symmetry tends to be more successful—a butterfly with asymmetrical wings can't fly very well, making it more prone to predation. Asymmetrical qualities are often caused by developmental challenges in the womb, which can include congenital defects, hormonal imbalances, poor nutrition, bad health, or substance abuse.

The preference for symmetry in mates isn't unique to humans; actually, the lack of a preference for symmetry would be uncommon. Symmetry is key to the mating success of many species, from the Japanese scorpion fly to the peafowl, from the zebra finch to the earwig.

Dr. Anthony Little made that connection when he discussed his team's study of attractiveness among British and Hadza people:

> Symmetry has been shown to be important in mate-choice in many animals. For example, female swallows prefer males with symmetrical tail feathers. While there may be cultural variation in preferences for other traits, we show that symmetry in faces is attractive across two very different cultures.

Symmetry isn't just an indicator of genetic health; it can also indicate reproductive health and fertility. As discussed in Chapter 1, women with evenly balanced breasts tend to be more fertile. Another small but interesting study showed that women involved with men who had more symmetrical bodies had more orgasms. Eighty-six women reported how frequently they had orgasms with their partners. The average was 60 percent. But among those with the most symmetrical partners, it was 75 percent; and among those with the least symmetrical partners, orgasm frequency dropped to 30 percent. Of course, this doesn't mean that body symmetry is actually *causing* more orgasms. It's more likely that body symmetry goes hand in hand with other qualities that combine to increase a man's overall attractiveness. "We don't think women are looking at asymmetry in hand width," observed Randy Thornhill, one of the leaders of the orgasm study, and a pioneer in human symmetry research. "Symmetrical males may be more dominant and have the highest self-esteem and this could influence their attractiveness."

The idea that at least some universal notion of beauty is connected to facial and body symmetry is not without controversy. And when coupled with the idea that it may be a marker of genetic and reproductive fitness, it can stir some pretty harsh criticism, especially in egalitarian societies that appropriately, teach us not to judge others on the basis of their looks.

But that won't stop Thornhill and others from searching for a possible connection. As he says, "Looks really matter. We're trying to find out what these looks are and how they evolved."

The more we learn and reflect about the hidden biological influences on our likes and desires, the more we're able to do something about it—regardless of sexual orientation to make up our own minds about whom we choose to spend our lives with.

FOR THOUSANDS OF years humans have turned to ornaments and body paint, makeup and all kinds of clothing, to make ourselves more attractive. And we've actually been going under the knife for the same reason for a lot longer than one might think—for centuries, in fact. Like so many other scientific breakthroughs that have later been put to more "recreational" (or at least optional) use, the first plastic surgeries were performed to meet a more urgent need. In the masterfully compiled book, *Aesthetic Surgery*, edited by Angelika Taschen, the contributors describe some of the first reconstructive surgeries, conducted in the sixteenth century to restore patients' appearance to some sort of normalcy after the ravages of syphilis. This sexually transmitted infection has a propensity to destroy the nose, leaving an obvious mark of "disrepute." Even before microbes were thought to cause disease, syphilis was understood to come from sexual vice. Unlike modern plastic surgeries—which are not always the most pleasant affairs—these early reconstructive procedures were downright excruciating, with multiple procedures extending over weeks and without any anesthesia.

Şerafeddin Sabuncuoğlu was a talented Turkish surgeon in the medieval Ottoman Empire who also happened to be a gifted artist. His two interests came together in *Cerrahiyet'ül Haniye*, or "Imperial Surgery," the first illustrated comprehensive guide to surgery—and one of the first to describe plastic surgeries of any kind. In it, Sabuncuoğlu provides a detailed description of mammary reductions performed to treat gynecomastia, the enlargement of breasts in males. He also discussed ambiguous genitalia at great length and techniques to repair *hypospadia*, a congenital malformation of the male urethra in which the urethra does not open at the normal place, but somewhere else on the head of the penis or even along the shaft.

All of these surgeries—and their modern successors—are known as reconstructive surgeries, one of the two types of plastic surgery, the other being cosmetic surgery, which is generally considered more optional. Strictly speaking, reconstructive surgery does not always reconstruct; rather, it corrects congenital defects as well as defects that occur later in life. It is surgery designed to create or restore proper function—a cleft palate or a hypospadia repair—while cosmetic surgery is more aesthetic. The distinctions can blur a little—a breast reconstruction after a mastectomy is considered reconstructive, not cosmetic, although it restores appearance, and not really function. The key difference is that, in the case of a breast reconstruction, for example, the surgery attempts to restore a woman's body to its "normal" symmetrical appearance; it is not designed to give her an appearance that is simply more desirable, which is what cosmetic surgery is about.

While plastic surgery has been around for many centuries, cosmetic surgery really has its roots in the last hundred years. As the authors describe in *Aesthetic Surgery*, many of the techniques used in cosmetic surgery today were first developed after World War I by German surgeons performing reconstructive surgeries on people disfigured in the war.

Hollywood started flirting with cosmetic surgery and dentistry almost as soon as it became available. Louis B. Mayer, the legendary studio boss, reportedly forced Greta Garbo to get her teeth fixed. Marlene Dietrich had a nose job—*rhinoplasty*—in 1929. And according to Norman Mailer, Marilyn Monroe had "bumps on the nose" removed and a "small flaw around the chin" corrected. The mole, we know, she left alone.

For a long time, cosmetic surgery remained the province of the very well-to-do or the well-connected, but in recent years that's been changing very rapidly. In 1997, Americans optionally subjected themselves to cosmetic procedures, both surgical and nonsurgical, 2.1 million times. According to the American Society for Aesthetic Plastic Surgery, that number almost quadrupled in the next six years, with 8.3 million cosmetic procedures performed in 2003—and a whopping 11.7 million in 2007. Those 2007 procedures cost $13.2 billion ($8.3 billion was for surgical procedures, and $4.7 billion was for nonsurgical procedures such as Botox injections)—making the quest for youth and "beauty" a very pricey practice.

It's not just Botox injections, nose jobs, and tummy tucks that are on the rise, either; less well known and less obvious cosmetic surgeries are growing in popularity too. One of those is labiaplasty, a procedure to reduce the size of the labia minora, the inner lips of the vulva. Most often, this two-hour surgical procedure is requested by women for cosmetic reasons, although some women report that the size or placement of their labia minora causes pain during intercourse.

Of course, there's no standard for "normal" labia minora, let alone an ideal. Just as eyes, ears, and noses come in enormous variety, so do sexual parts; from penises to nipples to vulvas, there's great range in size, shape, and coloring, and labia minora are no exception. Labia minora very rarely interfere with intercourse, but women with especially large ones can sometimes experience some discomfort. The general rule of thumb many doctors follow is that labia minora under two inches in width, usually don't need surgery for medical purposes. Of course, plenty of plastic surgeons are happy to perform cosmetic surgeries on labia minora of any size. Like other cosmetic surgeries, their take is, if it bothers you, then you can modify it. If you're wondering why women have labia minora, they may have served a somewhat useful function to keep foreign objects out of the vagina. Bear in mind that for most of human history we had no undergarments, or even clothing in general. Labia minora also become engorged with blood when a woman is aroused, which helps to *open* and facilitate entry into the vagina.

Size alone isn't the only reason some women choose to get labiaplasty. Just like breasts, it's common for one lip (labia minus) of the labia minora to be bigger than the other, and some women want them evened out—symmetry, again. Obviously, as with any cosmetic surgery, a well-informed adult ought to be able to opt for such a procedure in consultation with her doctor and others close to her, as she chooses. But the skyrocketing rate of cosmetic surgery does beg the question: when have we gone too far in pursuit of perceived physical ideals? Perhaps, when girls as young as ten years of age feel compelled to undergo labiaplasty to even out their labia minora.

IF THERE ARE some innate measures of attractiveness that people use in the hunt for a mate, it follows that more attractive people ought to have more success in the mating game. An Australian study confirms that's absolutely the case. Researchers from the University of Western Australia examined a large group of adults and looked for links between attractiveness and sexual activity. Here's what they found: the more attractive a man was, the more short-term sexual partners he had; the more attractive a woman was, the more long-term sexual partners she had. From a biological perspective, that can spell success.

Remember, for most of human existence, male reproductive success probably depended on finding many short-term sexual partners in the hopes that one or more of them would become pregnant, deliver a healthy baby, and raise it. Female reproductive success, on the other hand, depended on finding good physical traits when fertile, but then securing the best long-term partner to provide resources, protection, and stability in order to care for and raise children. As we discussed earlier, women are actually attracted to different types depending on where they are in their menstrual cycles. For the short period of peak fertility around the time of ovulation, they tend toward more high-testosterone, dominant, masculine types who might increase the chances that their babies will sire the most children themselves one day. The rest of the time, they find more feminine traits, like finer facial features more attractive, as they search for a long-term provider for their children. So, if being more attractive makes it easier for a man to find multiple female partners and a woman to find and hold the right male, attraction really does help to fix the reproductive sweepstakes in your gene pool's favor.

And a British study conducted back in 2001 garnered a lot of interest when researchers actually found evidence that looking an attractive person in the eye sparks activity in the *ventral striatum* that looking at less attractive people does not. The ventral striatum is an area of the brain that heats up in anticipation of a reward.

And, sure enough, brain scans of eight men and eight women shown a parade of images of forty different faces in four different ways revealed that the human ventral striatum is activated when someone looks into the eyes of an attractive person. The researchers concluded that the brain may consider the potential for social interaction with an attractive person a reward. Unattractive faces did not produce the same effect. Researcher Knut Kampe of University College London thinks catching the eye of an attractive person might activate the brain's reward center because we associate attraction with social status. "Meeting a potential good friend or someone who might influence our career might be very rewarding," he says.

Of course, the multibillion dollar pornography industry makes it clear that gazing into the eyes of beautiful people isn't the only kind of visual stimuli people find rewarding. Interestingly, despite the vastly greater number of men than women who buy and use pornography, a spate of new research indicates that women are aroused by visual stimuli just as easily and just as quickly as men. There are differences in the way men and women look at and respond to sexual images, but they're not necessarily what you'd guess. If I told you that one sex spends a lot more time looking at faces than genitals, while the other is aroused by a much broader range of sexual images, including fornicating monkeys, you'd probably guess that the first is women and the second is men. And you'd be wrong.

There has long been a consensus that men are more sexually responsive to visual stimuli than women. Dr. Daniel Amen, a psychiatrist and author of *Sex on the Brain*, even thinks that the way men respond to appearances is responsible for the cultural adaptation of makeup. "Most men are very visual, which is why women spend so much time on their appearances," Amen says. And there have certainly been studies confirming this, such as one by researchers at Emory University in 2004, which used brain imaging to reveal that key parts of the brain related to sexuality were more activated in men than they were in women when both looked at the same erotic images. But the situation is much more complex than that.

Another 2004 study, led by Meredith Chivers of the University of Toronto and the Center for Addiction and Mental Health, measured physical indicators of arousal in men and women who were shown a series of images. The women demonstrated clear signs of arousal in response to a greater variety of images—including pictures of bonobos (great apes closely related to chimpanzees) having sex—than the men did. The difference is that *the women didn't always know they were aroused*. Physical arousal, which can include vasodilation of the genitals, erections in men, lubrication of the vagina in women, and so forth, can precede a conscious sense of sexual desire, or mental arousal, the *thought* of sexual interest. In other words, your body can be aroused before you realize it.

Then, in 2007, researchers at McGill University measured arousal in men and women who watched pornography, by using thermal imaging to detect rising temperatures in the genitals. The rise in temperature was due to the increase in blood flow to the genital region, which normally prepares the body for sex. In the study, both men and women reached peak arousal in about ten minutes, casting further doubts on the widely held belief that women take longer to become aroused than men or are less responsive to pornography.

Also in 2007, Kim Wallen, one of the Emory professors who authored one of the earlier reports, teamed up with Heather Rupp, a fellow at the Kinsey Institute for Research in Sex, Gender and Reproduction at Indiana University, to produce a follow-up study. This time, the researchers not only used brain imaging to measure brain activity in response to sexual images; they also used eye-tracking technology to determine exactly what elements of the images their subjects focused on, and for how long. The combination allowed them to match brain activity to specific features in the images—faces, breasts, genitals.

Surprisingly, Rupp said, "Men looked at the female face much more than women, and both looked at the genitals comparably." Men were also more likely to look at faces first, and women were more likely to look longer at images of male-on-female sexual acts. The eye-tracking data offers an unexpected explanation for the more intense brain activity in response to visual stimuli men demonstrated in the 2004 study and others like it. Much of the increased activity is centered in the amygdala, which is deeply involved in processing emotion. So the increased brain activity may be the result of all the time men spend looking at *faces*. Men may be more consciously responsive to visual sexual stimuli than women because they're more emotional about it. The new vogue of today's sensitive man may not be so new after all.

By the way, paying for porn is no longer unique to humans. Researchers at Duke University offered male rhesus monkeys the chance to see pictures of female monkey bottoms, but only if they paid for it by giving up their fruit juice. The monkeys paid up.

■ ■ ■

IMAGINE FOR A second it's a beautiful spring day, and a woman sees a good-looking man across the aisle at the grocery store—you know . . . tall, dark, handsome, and symmetrical. She catches his eye, and she gets that familiar tingle. Why does that familiar tingle seem to be so familiar every spring?

Spring fever, of course. It's directly related to the increased sunlight that follows winter. The explanation for spring fever is actually pretty straightforward. Increased sunlight, which is detected by your eyes and via the optic nerve, and a group of neurons in the brain called the *suprachiasmatic nuclei*, eventually communicate to the part of the brain known as the pineal gland. The reporting of sunlight prompts your pineal gland to cut down its production of melatonin. Discovered in the late 1950s by a team of Yale researchers led by dermatologist Dr. Aaron B. Lerner, melatonin is a hormone that our body produces naturally which is involved in the regulation of circadian rhythms. This is the cycle of bodily chemistry and behavior that you follow from day to day— the most basic, of course, being awake and asleep. Melatonin has gotten some fair attention as an over-the-counter aid for long flights, supposedly helping people find midair sleep easier and jet lag less of an issue. And naturally occurring melatonin in our bodies is certainly linked to the desire to sleep and changes in mood. So, as we bask in spring's sunlit glow, we're also tamping down on the flow of melatonin, waking us up, lifting our mood, and, in many cases, possibly turning us on. Of course, after a long winter, the fact that it's finally spring doesn't hurt either.

Back to the guy across the grocery store aisle. She's now holding his gaze, and tilting her neck to one side. Her mouth is curling in a not terribly subtle hint of a smile. Is he staring right back at her, maybe raising his eyebrows a bit? Yes, they're flirting.

Flirting is the body language call and response of the mating game, and its vocabulary and grammar are deeply ingrained in our subconscious. "Flirting is a way of testing one's mate-value and the possibility of alternatives—actually trying to see if someone might be available as an alternative," says Arthur Aron, a psychology professor at the State University of New York at Stony Brook. When Irenaus Eibl Eibesfeldt filmed African tribes in the 1960s, he found women doing the exact same tilt of the head and little smile we just imagined a woman offering the man in the grocery store. Of course, flirting is not the sole domain of heterosexuals, everyone does it.

You see somebody who looks attractive. You flirt with them. They flirt back. At some point, you're probably close enough to smell them. You keep flirting. So do they. And whether it's another ten minutes, or over a series of dates, if you keep responding favorably to each other and things proceed, eventually it happens.

The first kiss. Why do we kiss in the first place? Zoologist and author Desmond Morris proposed in the 1960s that kissing might have evolved from primate behavior termed *prechewing*. This is the practice in which a mother would begin masticating or chewing food (prior to the modern convenience of commercial baby food), before passing it off to her young, using her mouth.

Regardless of how the practice of kissing arose, it is important to mention that it's not totally cross-cultural. In *The Expression of the Emotions in Man and Animals*, published in 1898, Charles Darwin wrote in reference to kissing: "It is replaced in various parts of the world, by the rubbing of noses." This behavior possibly refers to the practice of *kunik*, historically practiced by the Inuit and somewhat similarly by the Maori; it's part sniff, part nuzzle, not the commonly mistaken "rubbing of the noses."

In the majority of cultures where kissing is practiced, it can be a very important test of a relationship's potential. In fact, a study published in 2007 says the first kiss is so important it can kick a budding relationship into higher gear or cut it off altogether. In the study, 59 percent of the men and 66 percent of the women reported having been initially attracted to someone, but losing interest when the first kiss just didn't feel right. Why? The study's author thinks it's because you're still playing the mating game—gathering information, making judgments, assessing this person's suitability as a potential mate and possible partner—and you've just raised the stakes. And, when you kiss, you exchange all kinds of information with the person you're kissing. In an article in *Scientific American*, author George Gallup of the State University of New York at Albany said:

Kissing involves a very complicated exchange of information—olfactory information, tactile information and postural types of adjustments that may tap into underlying evolved and unconscious mechanisms that enable people to make determinations . . . about the degree to which they are genetically incompatible.

To put it in the most straightforward terms, if you kiss somebody who tastes unpleasant, it's likely to turn you off. Which makes me think that the odds are that bad taste means something; it could be a sign of a microbial infection—for example, the bacterium *Helicobacter pylori* that's associated with ulcers—or other parasites and even diseases. Or it might just mean all is not right.

For the most part, though, since we've made it this far as a species, nature has come to ensure that we enjoy sexual intimacy with others. Next we're going to find out how the way we look, smell, taste, and act come together in the big payoff: sex.

let's talk about sex

The experience of having sexual intercourse for the first time really runs the gamut—it can be unbelievably rewarding, utterly indifferent, or even emotionally damaging, depending on a host of factors, beginning with the age, maturity, and the relationship between the partners. But one thing it always is. It's new. And lots of cultures place enormous value on making sure that it really is new—that the first time is truly the first, especially for young women. In many of those cultures, people believe the only way to tell for sure is to examine the hymen. They believe that an intact hymen is the only true proof of female virginity.

The hymen is found directly at the vaginal opening. Hymens, like every other part of the body come in many different varieties. It's a mucous membrane that typically covers a portion of the vaginal opening in a circular shape or crescent.

It is estimated that one in every one thousand to ten thousand girls are born every year with an *imperforate hymen*—a hymen that covers the entire vaginal opening. If that doesn't resolve by puberty, it can prevent the flow of menstruation out of the body. Typically, young teenagers eventually show up at their doctors complaining of monthly pelvic pain and *primary amenorrhea*, which is the medical term for never having menstruated. An imperforate hymen needs to be opened surgically with a procedure called a *hymenotomy*, to allow the products of menstruation to flow out of the body. And, on the other end of the spectrum, although most girls are born with hymens, some are not. This can pose a real problem in societies where intact hymens have serious cultural significance. In fact, in some parts of the world being born without a hymen can be downright deadly. Among the Yungar people of Australia, for example, the lack of an intact hymen before marriage in the past, could have resulted in torture, forced starvation, and even death.

Through the ages women came up with some pretty ingenious methods to protect themselves from the consequences of a ruptured or absent hymen. According to Dr. Jelto Drenth, author of *The Origin of the World: Science and Fiction of the Vagina*, medieval Neapolitan women and nineteenth-century London brothel workers would actually use leeches to ensure a bloody first-night experience and thus convince their men of their virginity. The *Trotula*, a medieval medical text addressing the health complaints of women, gave the following advice about the use of leeches: "What is better is if the following is done one night before she is married: let her place leeches in the vagina (but take care that they do not go in too far) so that blood comes out and is converted into a little clot. And thus the man will be deceived by the effusion of blood."

Some women today still go to great lengths to convince their future husbands and in-laws, and even their own family, of their virginity. Sometimes, they even resort to surgery. These feigned virginity procedures can be done right before the wedding and include stitching the vaginal walls together (the stitches don't go in too deep) to ensure the presence of blood and a little resistance to insertion. *Hymenoplasty*, a procedure to restore a damaged hymen, is performed in many parts of the world, including Europe, where a growing segment of the population is composed of Muslims, who still place great importance on a bride's chastity. In many countries such as Turkey, performing a hymenoplasty is risky, both for the doctor and the patient, because there are laws and cultural traditions that punish performing or accepting such procedures. To convince doctors to perform a hymenoplasty without revealing the loss of their virginity, many women contrive stories that strain credulity, like falling on a fence. Even after finding a willing doctor, they still must come up with the money to pay for it. In the United States and Europe the procedure can cost anywhere from $2,000 to $4,000.

The hymen hasn't always been held in such high regard. Indian courtesans actively sought its removal. According to the *Kama Sutra*, thought to have been compiled sometime in the third century A.D., "the courtesan gets rid of her daughter's virginity with the assistance of a female friend or slave, so as to facilitate her amorous success. Once she has thoroughly studied the practice of sexual relations according to the *Kama Sutra*, she liberates her daughter. Such is the ancient custom."

For all the importance attached to them, are hymens the great indicators of virginity that they're thought to be? It turns out that even if hymens do tear, they can heal. And many women do not even have hymens—so if you don't have an intact hymen to start with, intercourse isn't going to tear it. It's also pretty darn easy to tear a hymen through nonsexual activity. Some women who are virgins will have already torn their hymens through some type of physical activity, or even tampon use. And, on top of all that, an intact hymen really isn't proof that you're a virgin. The hymen becomes very elastic at puberty, so much so that it can remain intact in many women even after intercourse. A study from the *Archives of Pediatric Adolescent Medicine* in 2004 found that half of the young women who admitted to having sexual intercourse actually still had intact hymens; in some cases, as hymens elasticize at puberty, they can stretch during intercourse without tearing. Some women have even become pregnant with intact hymens.

The bottom line is simple—a hymen might tear the first time, it might tear long before that, and, barring delivering a baby vaginally, it might not tear at all.

THERE'S PRETTY CLEAR evidence that there are some evolutionary pressures behind promiscuity, especially on the part of men (although men obviously don't have a monopoly on infidelity). To some degree, this explains why hymens may have been given such importance in so many different cultures. While evolution may have worked to encourage pair bonding and commitment through our chemical response to sex, it's not Cupid all the time.

Generally speaking, across most animal species, male reproductive success hinges on mass distribution, and female reproductive success depends on careful selection and conservation. Sperm are small, multitudinous, and continually replenished. Eggs are large, precious, and probably not produced after birth. In other words, sperm are cheap and eggs are expensive.

When a male and a female have sex, millions of sperm compete for the chance to fertilize, with very few exceptions, just one egg. And, if a given female has sex with more than one male, then the odds that any one of her individual male partners is going to succeed in reproducing drop accordingly, while hers increase. So, from an evolutionary perspective, the best way for a male to increase his odds of passing his genes on is to distribute his sperm as widely as possible among fertile females, in the hopes that sooner or later one of those sperm will beat the odds and fertilize an egg. Females, on the other hand, have an interest in seeking males who will give her the healthiest offspring so that the huge investment her body makes in pregnancy and child-rearing is worth it. Ultimately, males and females both have an interest in multiple sexual partners, but for different reasons.

And that's the way it is in most species; they're polygamous. But a few species—around 5 percent of mammals— are monogamous. More precisely, in scientific terms, they're known as socially monogamous, because they pair up with an individual of the opposite sex for life, although they are still very likely to mate outside the pair-bond.

In the last few years, scientists studying a furry little rodent called a vole have actually identified a single gene that is responsible for making one variety, the prairie vole, monogamous, while a closely related animal called the montane vole is polygamous.

Scientists have long known that prairie voles mate for life—couples live in the same nest, and males help to care for offspring; males, and sometimes females, may go off and copulate outside the pair-bond, but they always return. Male montane voles, on the other hand, are essentially all about one-night stands—there is no relationship after the mating act itself, even when it results in pregnancy and a litter.

The regulation of oxytocin and vasopressin (a closely related hormone), and their corresponding receptors in certain parts of the brain, is what is thought to turn a vole from frisky to loyal. Why would some voles evolve to be monogamous and others polygamous? Remember that evolution is all about trade-offs. Polygamy lends itself to genetic diversity and increases the likelihood that an individual male will pass his genes on. But monogamy creates a safer environment for offspring, giving them a better chance to survive, reach adulthood, and reproduce themselves, keeping the genetic chain alive. So in a particularly predator-heavy environment, for example, monogamy might give animals a greater advantage.

In that light, it's easy to imagine how competing pressures—for males to disseminate their sperm widely but also to protect their offspring—might lead to monogamous and polygamous tendencies in the same species.

Fascinating research, published in 2008, by Swedish geneticists has just linked the presence of a single gene that affects the location of vasopressin receptors in human males to the likelihood of marital problems. Remember, it's not just the *existence* of vasopressin that affects vole behavior—vasopressin exists in both species of voles; it's where that vasopressin is sensed, through the presence of their receptors, and processed in the brain. Hasse Walum, a behavioral geneticist at Sweden's Karolinska Institute, led a study published in the *Proceedings of the National Academy of Sciences* in 2008 that looked for copies of a gene called RS3 334 in 552 Swedish men who had been in a committed heterosexual relationship for at least five years. The team then interviewed the men and their partners about their relationships. The results were striking:

> Men with two copies of the allele [another word for a copy of a gene] had twice the risk of experiencing marital dysfunction, with a threat of divorce during the last year, compared to men carrying one or no copies. Women married to men with one or two copies of the allele scored lower on average on how satisfied they were with the relationship compared to women married to men with no copies.

So how prevalent is this gene? Forty percent of all men have at least one copy of it. But that doesn't mean that 40 percent of all men are going to have problems in their marriages. Genetic tendencies can influence our behavior, but we have the capacity to exercise control over impulses, especially when we understand the possible influences upon us. There are some people who can never imagine themselves staying faithful. They may even join a growing minority of couples opting for "open" marriages, where sex outside of the relationship is accepted. But, for the visible majority that choose the more traditional definition of marriage, knowing the triggers to infidelity can help tip the odds in your favor. As Helen Fisher, a biological anthropologist and the author of *Why We Love: The Nature and Chemistry of Romantic Love*, noted:

> There are many ways this information can help a man and his wife when they marry. Knowing there are biological weak links can help you overcome them. You can say, "Oh, it is just my DNA, and I am going to ignore it." . . . Some people will go into marriage with a stronger deck of cards. But there are people genetically prone to alcoholism who give up booze and make a good marriage. No one is saying biology is destiny.

Exactly. But sex is biology (and anatomy, anthropology, chemistry, psychology, and sociology, too). So let's delve a little deeper into what happens when men and women have sex.

LIKE JUST ABOUT anything else, sex gets better for most people with practice. The goal is good, rewarding sex with your partner, and that usually means one or more orgasms for the both of you.

So what exactly is going on when you have an orgasm? And why do we have them anyway?

Male and female orgasms are actually quite similar in many ways though female orgasms generally last much longer. But both involve a rapid series of muscular contractions in the genital and anal areas, every 0.8 seconds, in fact. Women may experience more intense contractions than men because the uterus can contract along with the vagina and the pelvic muscles. As both men and women experience orgasm, other muscles may shake or go into spasm. In some people, hands tighten, toes clench, and backs arch. Meanwhile, pleasure centers in the brain are highly activated, while activity in the cerebral cortex, where conscious thought takes place, momentarily recedes. It's no wonder the French call orgasm *la petite mort*, "the little death." The combination, of course, for most of us feels very, very good.

As to why we have orgasms, well, there's never been much disagreement over the evolutionary motivation for the physiology behind the delivery of genetic material from males: sperm pass up through the vas deferens into the ejaculatory ducts and then are combined with fluid from the seminal vesicles and the prostate itself to make semen. Contractions in the prostate and the penis force the semen into the urethra and out through the head of the penis. That is ejaculation. And that explosive generation of semen from the penis during vaginal intercourse is what launches sperm on their one-way journey to fertilize an egg. This has been the only way to achieve reproduction for all but the last thirty or so years of human existence.

From nature's point of view, the best strategy for male reproductive success is not too different from what some twentieth-century Chicago politicians supposedly advised their supporters when it came to voting: early and often. So in addition to the physiological evolution of the orgasm reflex as a means to release sperm, it's more than likely that the pleasurable nature of it created an impetus for men to seek out sex—and finish the job every time.

But what about women? Women can clearly conceive

without orgasm. So what's the evolutionary point behind female orgasm? That's a mystery about which the scientific community has still not come to agreement, although it's getting closer. The search began with the notion that just because female orgasm isn't required, that doesn't mean it doesn't help. As sex researcher Dr. Beverly Whipple, and her coauthors point out, in *The Science of Orgasm*: "An intricately coordinated physiological process between male and female sexual systems enables fertilization. In women, orgasm is evidently not one of the components that is *essential* to fertilization. . . . However, several studies . . . suggest that orgasm may *assist* the process."

Many theories have been advanced along those lines, some of which make sense, others that would be fine if they weren't so contradicted by the basic facts of typical human sexual interaction. For example, one outlandish theory argued that female orgasm served to tire women out so they would remain flat on their backs, giving sperm a chance to make their way past the cervix and into the uterus without leaking out. But as Dr. Elisabeth Lloyd pointed out in her 2005 book, *The Case of the Female Orgasm*, women are less likely than men to be tired or sedentary after orgasm, and they're actually much more likely to have an orgasm while on top. Which means if female orgasm is an adaptation to keep women lying still on their backs after sex, it's a pretty poor adaptation.

Another theory, advanced by Robin Baker, the author of *Sperm Wars*, and his colleague Mark Bellis, suggests that the contractions of the uterus during orgasm act to suck sperm into it through the cervix, aiding fertility. Baker and Bellis actually attached a fiber-optic camera to the base of a man's penis so they could film the intravaginal activity during female orgasm. And their footage certainly shows the woman's cervix repeatedly dipping into a pool of semen as her uterus contracts during orgasm, a phenomenon Baker calls "up-suck." Baker said the images this study delivered "completely changed my scientific understanding of what happens at the most critical moments during sex."

Of course, many disagree with this theory too, including Lloyd and Whipple, who both argue in their books that there are methodological flaws in the Baker-Bellis research.

Lloyd, in fact, argues against twenty different theories that suggest an evolutionary purpose behind female orgasm, concluding that the right theory is one first advanced by an anthropologist named Donald Symons in 1979: namely, that female orgasm is an accident. The argument is straightforward. Male orgasm is essential to male reproductive performance; female orgasm is simply the result of sharing the same basic wiring, left over from embryonic development during the first eight weeks of pregnancy, before sexual differentiation kicks in. "Females get the nerve pathways for orgasm by initially having the same body plan [as males]," says Lloyd. "Without a link to fertility or reproduction, [female] orgasm cannot be an adaptation." In other words, she seems to be arguing that if it doesn't help a woman get pregnant, it can't be a product of natural selection.

On the other hand, female orgasms are clearly implicated in forging stronger pair-bonds with men. As discussed earlier, women have better orgasms when they're in love, and more orgasms—so naturally I believe, regardless of sexual orientation, the possibility exists that orgasms in women could serve to strengthen the bonds that lead to love. Once again, evolution may have pushed sexuality in twin directions for men and women—male orgasms to encourage frequent sex and female orgasms to encourage attachment. Of course it's not so black and white. There exists a spectrum of gradations when it comes to everything in life, and so it is with human sexuality.

The bottom line is that the pleasurable payoff of sex may be more than just an end in itself. It may also be one of the means to creating enduring bonds between partners, driving people to become loving. "Recreational sex is thus supposed to function as the glue holding a human couple together while they cooperate in rearing their helpless baby," says Pulitzer Prize–winning author and physiologist Jared Diamond in his book, *Why Is Sex Fun?*

Remember how the chemistry of attraction may drive us to select genetically dissimilar, but genetically fit, mates in order to give our children a leg up in the genetic sweepstakes? Now, couple that with the notion that sex and orgasm help to create the kind of bonding with this new partner, a bond that you would otherwise only have with family members, and you can see how the combination may produce the outcome that gives us the best chance to have and raise children while coping with all the challenges of life. Instead of allowing the search for a prime genetic partner to isolate you, the resulting pair-bonding helps us to create a new family. And that not only helps us to survive and reproduce; it sweetens the process immensely.

Over time, regardless of gender or sexual orientation, the drive for sex can certainly change, but the bonding it helps to produce endures. Many couples, of course, are sexually active even very late in life (and the advent of drugs like Viagra has made that more true today than ever before), but conversation after conversation with long-term married couples has led me to conclude that it's not only great sex that keeps them together. Couples who are just as sexually active in their later years as they were when they first met are certainly the exception, not the rule. But the tenderness, affection, and abiding love that I have witnessed over and over again in hospitalized patients, as illness transforms one partner into patient and the other into caregiver, may have had some roots in sex.

One of the patients I cared for really brought home the power of love to help us endure hardship and suffering. He was an elderly man who had just lost his wife after fifty-seven years. This is what he told me:

> Sex with my wife was good, especially when we were young, and then, you know, the bills come, the kids come, and, well, at times it makes you wonder how your penis can get you into such a mess. But then at some point you actually realize that if it wasn't for that I would have missed out on so much in life, including my kids. And, boy, do I ever miss her.

It had been over two years since his wife passed away, and he still hadn't taken off his wedding ring.

GEORGE BURNS ONCE quipped, "Sex at age ninety is like trying to shoot pool with a rope." But that's changed for a lot of seniors, thanks in no small part to a little blue pill,

Viagra. Its discovery was sort of an accident. *Sildenafil citrate*, the generic name for Viagra, which recently celebrated its tenth anniversary, was originally developed as a treatment for heart disease. During clinical studies, men who were taking the drug repeatedly reported a surprising, and generally most welcome, side effect: they were getting serious erections. Drug giant Pfizer, sildenafil's developer, quickly realized that it had accidentally struck pharmaceutical gold and changed gears, bringing sildenafil to market as a treatment for erectile dysfunction. Recently, sildenafil has returned to its roots, as studies have shown it to be highly effective at treating a rare cardiovascular disorder called *pulmonary arterial hypertension*. Pfizer now actually sells the drug under two different brand names—*Viagra*, for erectile dysfunction, and *Revatio*, for pulmonary arterial hypertension.

With all the attention focused on Viagra's ability to dramatically improve the sex lives of seniors suffering from erectile dysfunction (not to mention their partners), not much focus has been on its other side effects. And one of them is pretty surprising: Viagra can cause serious nasal congestion, essentially by causing an erection . . . in your nose.

That's right, there's erectile tissue in your nose—the same type of tissue found in the penis and clitoris. The erectile tissue in the nose apparently regulates the intake of air between our two nostrils. Some of the latest thinking is that this allows us to smell in stereo, perhaps letting us better detect the direction a specific smell is coming from, in much the way that our brain calculates the direction of a sound by extrapolating from the difference in time between when the sound reaches each ear. This erectile tissue may also help to direct breathing to one side or the other when we are lying down to rest, ensuring that we get a full complement of air. If we lie on our right side, for example, perhaps partially obstructing airflow to the right nostril, the erectile tissue on that side will swell, ensuring that we get maximum air through our left—and unobstructed—nostril.

But in people who take Viagra, that erectile tissue can cause a surprising complication. Remember, nitric oxide triggers the dilation of those blood vessels so they flood the penis and fill its erectile tissue, causing it to expand and become hard. But when someone takes that little blue pill, it doesn't just interact with the blood vessels in the penis; it also causes the vessels that feed the erectile tissue in the nose to dilate. Normally, the erectile tissue in our noses works like a pair of pistons, keeping airflow stronger in one nostril, then the other. But when all of it swells under the influence of Viagra, it reduces airflow through the nose, often leaving patients with a feeling of congestion.

MEN AREN'T THE only ones who have sexually important erectile tissue. The clitoris and inner labia both experience increased blood flow when stimulated.

That increase in blood flow is an important part of the arousal process. As any woman who has explored her sexual responsiveness knows (and any man or woman who has explored it with her knows as well), female orgasms come in many different shapes and styles. They can be single or multiple, they can seem to teeter on a knife's edge for an excruciatingly long while, or they can explode all at once in a mad rush. And for some, they remain elusive. There are clitoral orgasms, which, for most women, are the most common type. There are Gräfenberg or G spot orgasms, described by many of the women who experience them as very different in intensity and character from a clitoral orgasm. There are blended orgasms, involving clitoral and G spot stimulation, there are orgasms from breast stimulation alone, and there are even mental orgasms, which can happen without any physical contact at all.

And there are ejaculatory orgasms. Yes, men are not the only half of humanity who can ejaculate. Women ejaculate too.

Now, if you're like many men and women today, female ejaculation may be news to you—but recognition of female ejaculation isn't actually new at all.

In his treatise *On Seed and the Nature of the Child*, Hippocrates explained his belief that the fluids released by women during sex were required, along with semen, to create life. The definitive Sanskrit text on love and sex, the *Kama Sutra*, says, "The semen of women continues to fall from the beginning of the sexual union to its end, in the same way as that of the man." Even twentieth-century American literature on sex and marriage, like the 1928 handbook, *Ideal Marriage: Its Physiology and Technique*, included descriptions of female ejaculation: "It appears that the majority of laymen believe that something is forcibly squirted (or propelled or extruded) or expelled from the woman's body in orgasm, and should happen normally, as in the man's case."

In the 1950s, as a wave of Puritanism flooded the United States, discussion of female ejaculation—and most matters sexual—disappeared. Female sexuality, in general, was relegated to the back burner, or shoved off the stove entirely. Men, their desires, and the satisfaction thereof were all that seemed to matter. Maxine Davis, author of a 1963 guide called *Sexual Responsibility in Marriage*, summed up the prevailing attitudes toward female sexuality when she asked, "Why all the hurrah about orgasm for women?"

Fortunately, the feminist movement declared a sexual revolution and demanded equality in the bedroom as well as the workplace, the voting booth, and the rest of society, reminding people what the authors of 1935's *Sex Practice in Marriage* understood: "No matter how tired a husband may feel at the moment, it is too unfair to obtain gratification from his wife without giving back to her, her own reward."

With a renewed focus on female sexuality and female orgasm, some people turned their attention to female ejaculation. The real groundbreaker was a paper by sexologists Beverly Whipple and John Perry published in 1981, reporting a case of one woman ejaculating. The paper described the woman being vaginally stimulated by her husband until she reached orgasm. And there, under the watchful eye of a team of researchers, she climaxed—and ejaculated, releasing noticeable amounts of fluid. As Dr. Whipple and her coauthors reported, "On one observed occasion, [the] expulsion was of sufficient force to create a series of wet spots covering a distance of more than a meter [over three feet]."

Today, the existence of female ejaculation is more accepted in the scientific community, although it still has its detractors. From survey results, some sexologists estimate that about 10 percent of women ejaculate during orgasm, although some put the number as high as 69 percent. In a 1988 study by Slovak researcher Dr. Milan Zaviačič, twenty-seven women were stimulated to the point of orgasm, and ten of them ejaculated, suggesting a prevalence of 37 percent. That study was followed by a 1,300-woman survey two years later in which 40 percent of the respondents indicated they experienced ejaculation during orgasm. I believe that the number of women who *can* ejaculate may be a lot higher. If that's the case, why don't more women experience ejaculation? It seems that the initial sensation, especially when the anterior wall of the vagina (the part of the vaginal canal that is just underneath the abdomen) is stimulated, is similar to the feeling women get when they need to pee. Not surprisingly, many women put a stop to the stimulation right there. They don't know it's leading to ejaculation, and they don't want to pee in the presence of their partners. Just like everything else we've discussed, when it comes to sexuality, variation is the rule. There are people who take part and enjoy urinating on their partners. It's called "water sports" or "golden showers."

The fluid produced during female ejaculation can be clear or milky white, almost like skim milk. It can range from a few drops to a quarter of a cup, and even more in some reported cases. Ernst Gräfenberg, the German doctor after whom the G spot is named, wrote about the potential scale of female ejaculation in a 1950 paper called, *The Role of Urethra in Female Orgasm*: "Occasionally the production of fluids is so profuse that a large towel has to be spread under the woman to prevent the bedsheets getting soiled."

In some women, it seeps out and is barely noticeable, contributing to the challenge scientists face in determining exactly how prevalent female ejaculation is. In other women, ejaculation can be incredibly forceful. One woman reported leaving "a patch of wetness some 2 feet in diameter." For some women, it happens once per orgasm; others can ejaculate over and over. One woman in Zaviačič's study ejaculated 160 times while under laboratory observation, with as little as thirty seconds of G spot stimulation.

Before Dr. Whipple's landmark report, many women who were ejaculating were wrongly diagnosed with urinary incontinence. Women obviously do produce liquid on a regular basis from their genitals—every time they pee. So, ignorant of female ejaculation, some doctors believed the source of all that liquid had to be the only source they were familiar with, the bladder. Essentially, they believed the ejaculatory liquid was a rush of urine produced by women who lost a little more control than others when their bodies were overcome with excitement during orgasm. Some women even underwent surgery in order to "cure" their incontinence.

Doctors aren't the only ones who mistake female ejaculation for urination, of course—and, in at least one case, that has made for pretty unlikely role reversals. In 2002, the British Board of Film Classification ordered six minutes and twelve seconds cut from a little nugget of high culture called *British Cum Queens*—and found themselves under attack by a group called Feminists Against Censorship. And what turned a small group of feminists into unexpected defenders of pornography? Female ejaculation education, of course.

The offending six plus minutes of video showed women producing liquid from their genitals. The film board said that was obviously golden showers, or urination, which is banned under Britain's Obscene Publications Act. That's not urination, it's ejaculation, responded the feminist group. Not necessarily, said the film board. The feminists presented a series of scientific studies in rebuttal, arguing strongly that female ejaculation exists. The board backed down from its claim, but maintained that it is a "controversial and much debated area." But it stood by its position that the scenes in question were illegal, "nothing other than straightforward scenes of urination masquerading as female ejaculation." In the film board's defense, most of the pornographic images that depict women ejaculating—often referred to as *gushing*—are faked using urine or a liquid inserted into the vagina prior to an orgasm or female ejaculation scene.

The board was certainly right about one thing—female ejaculation has been and still is the subject of much controversy. Many scientists today agree that it does exist, but it's taken a long time to get to this admission. Philadelphia gynecologist Dr. Martin Weisberg's first response to Whipple and Perry's report was: "Bull . . . I spend half my waking hours examining, cutting apart, putting together, removing or rearranging female reproductive organs. . . . Women don't ejaculate."

So Whipple and Perry set Weisberg up with a personal demonstration. Here's what he saw, in his own words:

The vulva and vagina were normal with no abnormal masses or spots. The urethra was normal. Everything was normal. She then had her partner stimulate her by inserting two fingers into the vagina and stroking along the urethra lengthwise. To our amazement, the area began to swell. It eventually became a firm one by two cm [0.4 by 0.8 inches] oval area distinctly different from the rest of the vagina. In a few moments the subject seemed to perform a Valsalva maneuver [bearing down as if starting to defecate] and seconds later several cc's of milky fluid shot out of the urethra.

Weisberg was a convert, but he didn't take it on faith, or even just trust his own eyes. The fluid was analyzed in the lab, and there were clear chemical indicators that set it apart from urine, including some that provided a direct link to male ejaculatory fluid. It carried chemical markers connecting it to prostatic fluid, the fluid released by the male prostate gland that makes up between 10 to 30 percent of semen. That's right, female ejaculate has a clear familial resemblance to prostatic fluid, chemically speaking, of course.

There's a good chance you've heard of *prostate-specific antigen*, or PSA, especially if you're a man over fifty or know someone who is. PSA is a protein made by the prostate gland that's present in semen and, at low levels, in the blood of healthy men. A high PSA blood level is a possible warning sign for prostate cancer, and doctors recommend that men over fifty get regularly screened for prostate cancer through a simple blood test. Elevated PSA levels don't always mean cancer. For example, benign prostatic hypertrophy (BPH) is a condition describing an enlarged prostate that can result in difficulty urinating and elevated blood levels of PSA. Why do men have PSA? The protein is a key chemical component of semen; it keeps semen from becoming too viscous, allowing sperm to swim freely. It is also thought to help dissolve mucus produced by the cervix, making it easier for sperm to make their way into the uterus.

Sure enough, when scientists conducted chemical analyses of female ejaculate in the laboratory, they found significant levels of PSA, as well as PAP, or prostatic acid phosphatase, an enzyme produced by the prostate gland. What they didn't find was just as telling—the levels of urea and creatinine, the two main chemical signposts in urine, were very low, far lower than they would be if the liquid were urine. And finally they found glucose and fructose, two natural sugars. This could explain why people who have tasted female ejaculate have reported that it tends to be sweet. Fructose is also a key ingredient in semen, where it provides energy for sperm.

The most recent study of female ejaculate, published in 2007, was led by the Austrian urologist Florian Wimpissinger and his colleagues at Rudolfstiftung Hospital in Vienna. They examined two healthy women in their midforties who regularly reported ejaculating during orgasm. The liquid produced by these women was subjected to biochemical analysis, and the scientists also used sonography to image the female prostate of their subjects. The results were conclusive:

> Biochemically, parameters of the examination of the fluid emitted were clearly different than urine voided prior to sexual activity. Biochemical parameters—with special reference to prostate-specific antigen (PSA) . . . show that the source of fluid expulsion during orgasm is not urine, but is rather similar to male ejaculate.

So where do all these prostatic compounds in female ejaculate come from? The female prostate, of course.

Reinier De Graaf, a seventeenth-century Dutch physician, made a series of discoveries in reproductive anatomy before his death at thirty-two, one of which modern medicine has only now returned to. In 1672, he documented a collection of glands and ducts around the female urethra that produces a "pituitoserous juice" that makes "women more libidinous with its pungency and saltiness and lubricates their sexual parts in agreeable fashion during coitus."

According to Dr. Catherine Blackledge, author of *The Story of V*, prior to the 1880s, "it was generally accepted that women had a prostate too." It was the Scottish born gynecologist, Alexander Skene who "chose to focus on just two of the many glands of the female prostate," writes Blackledge. The Skene's glands (also called the para-urethral glands) are thought to drain their fluid through two pinhole-sized openings just above the vagina. In 2001, after examining more than 250 peer-reviewed scientific studies, the Federative International Committee on Anatomical Terminology (the official namer of names when it comes to human body parts; as they say, "the only internationally accepted source for human anatomical terminology") officially renamed the Skene's glands and the mass of tissue behind them that surrounds the urethra as the "female prostate."

In human males, the prostate is usually around the size of a walnut, weighing about four-fifths of an ounce. In men, the prostate surrounds the urethra like a doughnut. The female prostate, which surrounds the urethra, seems to come in a variety of shapes and sizes.

Sex educator Deborah Sundahl describes those shapes and sizes in her book *Female Ejaculation and the G-Spot*, using research compiled by Milan Zaviačič, the Slovakian pathologist. According to his research, the great majority of women, around 70 percent, have a ramp-shaped prostate along the urethra with the thickest part near its opening. About 15 percent have the reverse, with the thickest part at the far end of the urethra, near the bladder. A smaller number, about 7 percent, have a prostate that is thickest in the middle, and about 8 percent of women have what he called a "rudimentary" prostate, with very few glands.

In the Wimpissinger study of the two women who regularly ejaculated, high-definition sonography revealed "A hyperintense structure surrounding the entire length of the urethra with the anterior wall of the vagina adjacent. . . . It closely resembles that of the male prostate."

For those who believe in its existence, the G spot is located in the area of the vagina along the upper wall, but the precise location is said to vary from woman to woman. And the collection of glands and ducts that make up the female prostate run along the urethra on the same side of the upper wall of the vagina.

To many sex researchers, it's beginning to look as if the G spot is actually the best spot in any given woman to stimulate her prostate.

In men, the prostate contributes some of the fluid that is present in male ejaculate. When some women become aroused, the anterior part of their vagina (behind which is female prostatic tissue) increases in size, exposing an erogenous zone that is otherwise less prominent. Given its placement, it's easy to see that the best place to stimulate it would be the area on the anterior or upper wall of the vaginal canal where it's easiest to access. As an anecdotal aside: some men can be brought to orgasm through prostate stimulation, and they describe an increase in intensity very similar to the terms women use to describe a G spot orgasm.

All told, it's been estimated that about 90 percent of women may have a prostate. It's quite possible that the range in size and location of the female prostate may contribute to the ease and frequency of female ejaculation.

The male prostate is entirely encased by a fibromuscular sheet made of smooth muscle cells that, when they contract, help to expel prostatic fluid into the urethra, where it mixes with other seminal fluid just before ejaculation. It's not clear whether the female prostate is surrounded by similar tissue; but, in the same way that variation in size and location may contribute to a woman's ability to ejaculate, it's possible that the relative scarcity—or abundance—of these cells may account for the range of explosive force women experience when ejaculating.

One thing is sure. As Dr. Wimpissinger's report concludes: "Female ejaculation—first described as 'love juice' in ancient Indian textbooks—seems to be more common than generally recognized."

It's entirely a personal choice, of course, but if you want to experiment with female ejaculation, there are all kinds of guides and courses available. When I spoke with Whipple, she stressed the importance of encouraging women to enjoy what they find pleasurable and not to be set on finding the G spot or experiencing female ejaculations as the only goal. Her work has been to validate women's experiences not to set new goals. Given that, she describes the best way to start:

Begin with clitoral stimulation, and at least initially this may be the best way to start getting aroused. One should never rush or feel pressured in any way. When she is ready it's best to move on to stimulating the anterior wall of the vagina [which, if a woman is lying on her back, would be closest to her stomach], stimulating that area with one or two fingers making a "come here" motion. It's perfectly normal if she feels as if she needs to urinate, because this area surrounds the urethra, the tube you urinate through. With time and practice and help of a partner a woman may experience female ejaculation.

If a woman wants to experience ejaculation, she has to let go and allow it to happen, which requires letting go of a lifetime of training when it comes to bodily fluids in bed. When I spoke to Deborah Sundahl about her experiences teaching hundreds of women to ejaculate for the first time, she said, "The biggest obstacle to women ejaculating is not letting go. That is also men's biggest complaint about women sexually— they won't let go. If they do they're going to ejaculate all over you and that could be taboo."

THE EVOLUTIONARY RATIONALES behind the male prostate and ejaculation are pretty clear. Male prostatic fluid helps to ensure that semen is the right viscosity for sperm to swim easily, and it may also help them to clear the cervix by thinning cervical mucus. And ejaculation, of course, is what sends sperm on their one-way swimming race. But is there a purpose for female ejaculation? Is the female prostate a gland with a mission or a biological leftover from those early stages of common male and female embryonic development, not unlike the male nipple?

The answer to why our bodies would produce a fluid ought to be found in what that fluid does. Let's start by taking a look at the male prostate and prostatic fluid, which has been studied extensively, for some clues into the nature of the female prostate and prostatic fluid.

The prostate has the highest concentration of zinc in the body, and prostatic fluid is exceptionally rich in zinc. So what does zinc do, you ask? Well, it does a lot, but one of the main things it does is cause trouble for bacteria. Bacteria feed on iron, which they get from the organisms they've infected, including ours. Chemically, zinc is something of an iron mimic, which can confuse bacteria, gumming up their cellular machinery and making zinc a potent antimicrobial. It's so effective that it's used in commercial products—millions of mothers use Penaten baby cream to treat diaper rash. Penaten is thought to be so effective because it's 18 percent zinc.

Now, zinc isn't the only compound in prostatic fluid that has a family connection to the microbe-fighting business, so let's follow this thread for a moment. Prostasomes are small, vesicle-like structures with multiple functions, many related to fertilization. One of their principal roles is to protect sperm and improve their ability to swim. Prostasomes are coated with a chemical compound hCAP-18, which is thought to be produced by cells in the testes. The compound hCAP-18 is the chemical precursor (or forerunner) of a powerful antimicrobial compound called LL-37, which is found in many places throughout the body.

So why would women's urethras need an antimicrobial wash, with compounds not normally found in urine, every time they ejaculate? To prevent urinary tract infections, of course.

Urinary tract infections are incredibly common in women, much more common than in men. Half of all women will develop at least one urinary tract infection at some point in their lives, and the risk of recurrence increases with each subsequent infection. For many women, they are a recurring, painful, fact of life: 20 percent of the women who have a first urinary tract infection will have a second; 30 percent of those who have a second infection have a third; and for those unlucky women, 80 percent have additional recurrences in store.

Why are women so much more likely to suffer from urinary tract infections than men? It probably comes down to anatomical geography. The female urethra, with its opening just above the vagina, is close to bacteria that normally reside there, making it easier for them to get in.

Female urinary tract infections are so common that they've given rise to a host of myths, about both what causes them and what cures them. Women have been told that failure to pee after sex, and even failure to douche, can lead to an increased risk of infection. But study after study has shown that one behavior widely thought to increase the risk of recurrent urinary tract infections does exactly that—sexual intercourse. There's a pretty logical explanation for how that could occur, of course, and, rest assured, it's nothing specific to your partner. Although the possibility does exist that some partners may harbor strains of certain microbes that are more adept at causing urinary tract infections. What seems to happen is that the actual act of intercourse, especially if it involves changing positions during sex, can introduce and move microbes into the urethra, where they can even make their way into the bladder.

There are other behavioral risk factors for recurrent infections. Studies have shown that using a diaphragm and certain spermicides can also increase risk, so if you have a problem with urinary tract infections and use either type of product, be sure to talk to your doctor about it. And new research has shown that there is probably a genetic component to risk, involving the susceptibility of the wall of your bladder to colonization and infection.

Still, an active sex life is definitely one of the most common risk factors. As one sufferer said to Dr. Judith Reichman, a physician who specializes in women's health issues, "I seem to get bladder infections every time I have sex. What would you suggest? (I hope it's not abstinence!)"

I too hope it's not abstinence—and it doesn't have to be. But there certainly does seem to be merit in doctors' common nickname for recurrent urinary tract infections, or UTIs. They call it "honeymoon cystitis," and it can certainly wreck at least *that* part of your honeymoon. Urinary tract infections definitely get in the way of sex. "Obviously, it is difficult to enjoy sex with excruciating suprapubic pain," writes clinical psychologist Naomi McCormack in the journal *Sexuality and Disability*. "One of the most popular sexual self-care strategies is to avoid sexual activity, especially intercourse."

If a single episode of sexual intercourse would cause an infection that made women avoid sex, wouldn't evolution do something about it?

That's exactly what I think happened. It turns out I'm not the only one who came to this conclusion. Whipple and Perry hypothesized back in 1982 that female ejaculation may have evolved to help prevent urinary tract infections in sexually active women. New research needs to be done, and that's why I've begun to investigate this very issue. It's not yet known, for example, whether zinc, LL-37, or any other antimicrobial compounds are present in female ejaculate in appreciable amounts, as they are in male ejaculate. But, if studies demonstrate a significant antimicrobial character in female ejaculatory fluid, and I believe they will, it would be strong evidence that female ejaculation has a real purpose.

Of course, if you *do* suffer from recurrent urinary tract infections, ejaculation isn't your only hope. Infections can be treated with antibiotics. But the old tale about peeing after sex to clean out bacteria won't help—numerous studies have shown that regular urine does nothing to decrease the risk. There is one folk remedy that shows signs of standing up under scientific scrutiny, at least for women who suffer from bouts of symptomatic recurrent UTIs. Cranberry juice.

Professor Itzhak Ofek of Tel Aviv University has been studying cranberries and their healing properties for more than twenty years. He and his colleagues published a widely disseminated report in the *New England Journal of Medicine* that documented their identification of a molecule in cranberries called nondialyzable material, or NDM.

NDM basically acts as a shield for some cells, coating them and insulating them from infection by preventing microbes from attaching and setting up shop. Professor Ofek says, "We understood that there was something in cranberry juice that doesn't let infections adhere to a woman's bladder. We figured it was a specific inhibitor and proved this to be the case." Not to discount Professor Ofek's research, but it's at least worth noting that a majority of his research was funded by Ocean Spray, the cranberry juice giant.

Speaking of legends and folk remedies, the idea that oysters are an aphrodisiac is said to have first sprung from the lips of legendary lover and sometime author Giacomo Girolamo Casanova de Seingalt, or just plain Casanova for short. Guess what food contains more zinc per serving than any other? Oysters.

Whether you find your zinc in your ejaculate or in your oysters, it's bound to help keep the microbes away.

IF FEMALE EJACULATE helps to keep women healthy, it may not be the only bodily fluid that lends a hidden helping hand. Semen may help to get women pregnant—which sounds obvious—but not just by supplying sperm. In 2003, an email circulated that included what appeared to be a CNN report discussing a study that linked women giving oral sex to a decrease in breast cancer. There was no such study—just a well-crafted college student joke. The email led to the embarrassment of some international media outlets that took it seriously and ran with it. Even though semen exposure through oral sex doesn't decrease breast cancer risk, it may very well increase the chances of a successful pregnancy.

Here's the idea in a nutshell. In women's bodies, sperm are foreign entities, which means they should be treated like other foreign microbes—attacked and destroyed by the immune system. That doesn't happen, at least not all the time—if it did, of course, we wouldn't be here today. But it may be that a woman's immune response to *invading* sperm is yet another hurdle sperm have to overcome in their race to fertilize an egg. It's possible that repeated exposure to a partner's semen (and a chemical called TGF-beta and the sperm within) may *familiarize* a woman's body with it and make it more acceptable. If familiar sperm are treated more favorably than other sperm, it might provide another advantage as well: it would decrease the odds of pregnancy from a sexual encounter with a stranger, while increasing the odds for reproduction with a long-term mate. This theory is supported by new research that found that women who engaged in oral sex with their male partners were less likely to suffer from a rather dangerous pregnancy-related hypertensive crisis called preeclampsia. This condition can occur during pregnancy (family history, especially in sisters, is just one of the risk factors), in which blood pressure rises to potentially dangerous levels. Some women with preeclampsia can even deteriorate further, developing eclampsia, which is characterized by seizures and classified as a true obstetrical emergency. Providing even stronger evidence that the link between oral semen exposure reduced the risk of preeclampsia, the women who reported swallowing their partners' semen had the most protection.

And, of course, the key may very well be overall semen exposure—not just oral sex. In all associative studies, while trying to tease out relationships between different factors, in this case semen exposure and preeclampsia, it's difficult to know what is truly causal. For example, women who engage in oral sex and swallow semen have been found to be more likely to engage in other sexual behaviors more frequently than women who don't, including unprotected vaginal sex and anal intercourse, both of which could increase their exposure to semen. And though this research looks interesting, so far the results are not confirmed. Suffice it to say, before doctors begin recommending oral sex and semen exposure to women trying to conceive, *much* more research still needs to be done.

But if you look at all the research from on high, the trend seems to confirm something humans have thought for millions of years—sex can be pretty darn good for you.

come as you are

Nothing in evolution is free. Every adaptation is a trade-off, a compromise of some sort: walking on two legs gave us the advantage of height but cost us in terms of speed; our bigger brains give us the advantage of higher intelligence, but much of that brain growth occurs after birth, which makes human newborns especially helpless. In that sense, nature is really a massive arbitrage that uses trial and error to arrive at certain biological traits through natural selection that, across a given species, confers more benefits than costs. And sexual reproduction is no exception—it comes with significant costs, although it clearly has benefits.

Here's the general thinking about the overall benefit of sexual reproduction: sex makes a species more flexible by reshuffling the genetic deck with every generation, as well as purging parasites in parents instead of passing them on. Flexibility means that a species has better odds of finding the right adaptation to survive environmental changes that bring new threats in the form of microbes, climate changes, new predators, and so forth. In fact, some organisms, such as certain types of algae, can reproduce asexually and sexually. When it comes to "choosing" the type of reproduction, if there's a serious enough change, they choose sex.

One of the best explications of this theory is the Red Queen hypothesis, which was popularized in the excellent book, *The Red Queen: Sex and the Evolution of Human Nature*, by Matt Ridley. The term was first coined by evolutionary biologist Leigh Van Valen of the University of Chicago to describe how species turn to sexual reproduction in order to survive in a changing environment in which other species are constantly evolving too. Van Valen got the name from Lewis Carroll's book, *Through the Looking Glass*, the sequel to *Alice's Adventures in Wonderland*. In the book, Alice is running in a race but getting nowhere, when the Red Queen explains to her, "It takes all the running you can do, to keep in the same place." In other words, in a world where all manner of living things— from bacteria to plants to predators and prey—are evolving themselves, it takes a lot of evolution to keep up with the competition and thrive as a species. If a species stops evolving, it falls behind. Sexual reproduction gives a species the chance to undergo evolutionary experimentation every generation—it's the running a species needs to do to keep in the same place.

But, as I said, sex is costly. Chemistry and biology are like every other process on Earth—the more complex a routine, the more room there is for error. The asexual reproduction of a single-cell creature through simple cell division is, at least on the surface, pretty straightforward, biologically speaking. It's reproduction through carbon copy, very efficient and somewhat resistant to error. Sexual reproduction is the polar opposite—it uses enormous resources, requires two parents instead of just one, and is rife with possibilities for error as genes from two distinct organisms are blended together to make a third. In other words, sex is expensive.

So what exactly are the costs, especially to us? And if it's so expensive, why is sexual reproduction so popular from an evolutionary perspective?

THE FIRST COST of sexual reproduction is the possibility that something can go wrong from the get-go, during development. Sexual and gender differentiation in humans is a massively complex choreography of chemistry and biology that transforms an undifferentiated seven-week-old embryo into a male or female fetus and ultimately, if everything goes according to normative development, a baby boy or a baby girl. The one thing you can always count on with a biological process as complicated as this one is that things don't always go according to plan.

The truth is that developmental complications in sexual differentiation are far more common than most people realize. The Intersex Society of North America (ISNA), one of the major advocacy groups for people with sexual development disorders, put the rate of ambiguous genitalia at about one in two thousand—but that rate may far underestimate the true incidence of sexual development disorders that occur, especially if you include the number of naturally occurring aborted pregnancies that don't continue (usually with a high number of genetic and physical abnormalities).

A controversial study in 2000 by Brown University researcher Anne Fausto-Sterling used a comprehensive examination of the medical literature from 1955 to 1998 to estimate the frequency of the complete range of sexual variation disorders. The report concluded that one out of every one hundred people has a body that differs somewhat from the standard male or female configuration.

A note on terminology, before we go any further. Sex can be defined in many ways. The most basic way is to use chromosomes. If you have a Y chromosome, then you are considered male, at least genetically. The traditional term for an individual with both male and female reproductive organs is *hermaphrodite*, but true hermaphrodites—people with both ovarian and testicular tissue—are extremely rare. Female pseudohermaphroditism occurs when a child is born with the normal 46 chromosomes and two X chromosomes (46-XX; i.e., genetically female) but with ambiguous or underdeveloped female genitalia. Male pseudohermaphroditism occurs when a child is born with 46-XY (genetically male) but has the external physical appearance of a female, a lack of virilization usually resulting from insufficient testosterone production or insensitivity to testosterone. There is a third category of conditions that occurs when a child does not have the normal complement of chromosomes; from a medical perspective, these conditions are considered more in the realm of birth defects. We'll discuss conditions in all three categories in this chapter.

As medical science has become more sensitive to and

thoughtful about these conditions, the language used to discuss them has evolved as well, moving somewhat away from older terms, like hermaphrodite that are not particularly descriptive from a medical perspective and have become laden with pejorative connotation. At the 2005 Intersex Consensus Meeting, attendees agreed to adopt the term *disorder of sexual development*, or DSD, to cover the wide range of disorders that prevent an individual from being identified as typically male or typically female. The whole idea behind moving toward a clinical-sounding phrase was exactly that—to move toward a more scientific approach. Along those lines, the term *intersex* itself has given way for some to DSD as well, although groups like ISNA haven't changed their names. Disagreement within the intersex community still exists about whether the name change is helpful. Cheryl Chase, executive director of ISNA, told *Scientific American* it's her hope that the name change will encourage doctors to see DSDs as lifelong medical conditions. "Now that we've accomplished the name change, culture can accomplish a little magic for us."

IN DECEMBER 2006, Santhi Soundarajan's career as an elite athlete was about to really hit its stride. She had just won the silver medal in the women's 800-meter race at the Asian Games in Doha, Qatar, and seemed destined for the 2008 Olympic Games in China. And then, just hours before she was supposed to be honored, her silver medal was taken from her, and she was barred from further competition as a member of the Indian team by the Athletics Federation of India. For doping? Some other form of cheating? No.

She failed a sex test. Officials wouldn't say exactly what the tests showed, but according to the online science magazine *Inkling*, some anonymous official told the Associated Press that Soundarajan had "more Y chromosomes than allowed." Of course, one Y chromosome is all it usually takes to make someone genetically male. Here's the thing, though: Soundarajan had apparently passed sex exams many times before. So what happened?

The exact nature of Soundarajan's prior tests and the findings of the test that ultimately disqualified her aren't publicly known, but we can hazard a few good guesses. First of all, it's quite possible that Soundarajan's initial exams were simply physical inspections of her genitals, and that she possesses genitals that look sufficiently female to pass such a test. Fears of men masquerading as women in athletic competition are not unfounded, although as far as anyone knows, they are extremely rare. One of the few times it's actually thought to have occurred was during the 1936 Berlin Games when German Hermann Ratjen competed as Dora in the high jump. Funny thing is, he failed to place.

Soundarajan wouldn't be the first athlete to pass a sex test and later run into trouble because of a genetic analysis of chromosome makeup that discovered a Y chromosome. The story of Spanish hurdler Maria José Martinez-Patiño became well known when she ran into problems in 1985 with the discovery that she carried a Y chromosome.

Here's a quick primer on how sexual differentiation works. For seven weeks after conception, there's really no observable difference between genetic male or genetic female embryos. Then, in a normal male fetus, one with a single X and a single

Y chromosome, the gene on the Y chromosome, called the sex-determining region Y, flips on and produces a protein that causes the gonads of the fetus to transform into testicles. If SRY isn't present at all, or the genetic sequence is damaged, then the gonads become ovaries. Of course, that's what we know so far. The picture of sexual development is far from completely understood, and scientists believe many other genes are involved in orchestrating the process of testicular development alone.

Once the testicles are formed, they begin to produce testosterone. Testosterone and other sex hormones, such as dihydrotesterone (DHT), that spur the development of male sexual characteristics are collectively called androgens. In the womb, androgens trigger the development of the penis, the scrotum, and the descent of the testicles, along with the requisite cardiovascular connections of arteries and veins, from the abdominal cavity. At puberty, androgens are responsible for the growth of sperm and the development of male secondary sexual characteristics—male traits like dense pubic hair, facial hair, and increased muscle mass.

To trigger those developmental changes, androgens must bind with specialized receptors. Sometimes, even though an individual has what appears to be a standard genetic male chromosomal XY pattern, or karyotype, a genetic anomaly produces faulty androgen receptors. When that happens, the androgens can't attach to their receptors and, as a result, they have little or no effect, depending on the exact nature of the anomaly. This is called androgen insensitivity syndrome, or AIS; it's thought to occur as frequently as one in twenty thousand live births.

In a person with AIS, the SRY gene triggers the development of testicles as normal, but all the androgen-triggered changes just don't happen. The genital ridge does not become a penis and scrotum and the testicles never descend. And, since the "default" configuration of an embryo is female, the baby is born with a clitoris, vaginal labia, and a vaginal opening. Although at the moment medical science considers the female configuration default, there may be other factors that we have yet to discover that determine the exact intricacies of differentiation. At the same time, at six to seven weeks, the testicles still produce anti-Mullerian hormone, that initiates the disintegration of the Mullerian ducts, the precursor to the rest of the female reproductive system. Anti-Mullerian hormone is not an androgen, so, like the testis-determining factor produced by the SRY gene which prompted the transformation of the gonads into testicles, its role in the developmental process is unaffected in these babies. The result is an infant who looks just like a normal baby girl on the outside; but internally the picture is very different. The vaginal canal is shorter than normal and can end in a blind pouch. There is no cervix or uterus, there are no Fallopian tubes, and instead of ovaries, there is an internal pair of undescended testicles.

At puberty, the testicles in such cases produce increased levels of testosterone as they normally would in a male. Some of that is transformed into estrogen, again, as it normally would be. But in a person with AIS, testosterone has limited or no effect, since cells don't respond to it. It's important to remember that development of some secondary female sexual characteristics does not always need ovaries and female sex hormones. The typical adult with AIS, despite being genetically male, has female breasts, wide hips in keeping with feminine fat distribution, no extensive facial hair, and sometimes less pubic hair. And because there is no uterus and no ovaries, there is no menarche. At the same time, the internal male organs, like the epididymis, vas deferens, and seminal vesicles, have not developed because they all depend on testosterone to do so.

Sometimes parents learn of AIS when a predelivery genetic test, like chorionic villus sampling (CVS) or amniocentesis, reveals a male chromosomal pattern that does not match images on the sonogram (sex characteristics can sometimes be observable as early as nine weeks). In the absence of such a genetic clue, AIS may not be diagnosed until a concerned teenager heads to the doctor as she begins to wonder why she hasn't had her first period. And of course, sometimes AIS isn't diagnosed at all.

In Maria José Martinez-Patiño's case, her eventual diagnosis with AIS simultaneously explained how she could have a Y chromosome but otherwise have lived her life entirely as a female. Why didn't anyone ever think twice about the fact that Maria didn't menstruate? Remember, we discussed that menstruation is linked to body fat. If a woman's body doesn't have sufficient fat to support a pregnancy, ovulation and, thus, menstruation are suppressed. This evolutionary mechanism prevents women from having children in times of famine or poor food supply, better to survive hard times and reproduce later than to lose both mother and child. Elite female athletes, especially runners, often have such low percentages of body fat that they experience menstrual suppression—so Maria may never have thought anything was out of the ordinary.

Humans normally have 23 pairs of chromosomes, for a total of 46. They are all matched pairs, each a copy of the other, except at times for the twenty-third pair, the sex chromosomes. When chromosomes come in pairs, it's called *disomy*. Sometimes, because of a reproductive error, a human embryo might have three copies of a chromosome, or *trisomy*. Most embryos that have trisomy don't survive, but sometimes they do. Down syndrome occurs when there is trisomy of chromosomes 21. A single copy of a chromosome is called *monosomy*. Most human embryos with full monosomy don't survive; one exception is monosomy of the X-chromosome, when a woman is only born with one X-chromosome, resulting in Turner syndrome, a condition we'll discuss shortly.

Women can also be born with an extra X chromosome; triple X syndrome is thought to occur in one out of a thousand newborn girls in the United States. Many women with triple X syndrome don't have any symptoms, but when they do they can include above-average height or developmental delay. Klinefelter syndrome occurs in males (and males only) who have an extra X-chromosome, making them genetically XXY instead of XY. If a boy with Klinefelter syndrome has one extra X chromosome, he'll have a total of 47 chromosomes instead of the normal 46. This is often referred to as 47-XXY, where 47 indicates the total number of chromosomes and XXY indicates the karyotype of chromosome 23. Genetic males can also have XYY syndrome; instead of having an extra X chromosome as in Klinefelter's, men with this condition have an extra Y. As with triple X syndrome, they don't usually exhibit symptoms except for increased height and are thought to be at higher risk for learning disabilities.

Klinefelter's isn't limited to trisomy; boys with it can actually have more than just one extra X chromosome. Tetrasomy

(XXXXY) and pentasomy (XXXXXY) have been documented; as the number of extra X chromosomes climbs, the more pronounced the symptoms of this condition tend to become. One of the reasons that a child with tetrasomy or pentasomy can survive is due to X inactivation: a special feature that comes with having more than one X chromosome, extra copies can shut themselves off.

Although the exact prevalence isn't known, Klinefelter syndrome is thought to occur in about one in every five hundred to a thousand males, and is the second most common genetic disorder overall (in live births) that is caused by extra chromosomes. (Down syndrome is the most common.) Men with Klinefelter's are almost always sterile. Physically, they can be characterized by having small testicles, are often relatively tall, exhibit a tendency toward higher levels of body fat, and are more likely than normal XY men to have gynecomastia, which is the development of female-looking breasts.

Turner syndrome is a disorder in females caused by a missing chromosome, a monosomy of the X chromosome: instead of the normal XX, females with Turner's are just XO. This, of course, gives them a total of 45 chromosomes. A normal female karyotype is 46-XX; a woman with Turner syndrome is 45-X.

Because they have only one X chromosome, girls and women with Turner syndrome generally produce insufficient quantities of various hormones, including estrogen. These hormonal deficiencies affect the sexual and physical development of women with this condition. They are almost always infertile, although advances in fertility therapy have allowed some women with Turner syndrome to become impregnated with an embryo created from a donor egg and carry a pregnancy to term. Outwardly, Turner syndrome is often characterized by short stature, minimal breast development, and webbing of the neck. This condition also correlates with an increased risk of cardiovascular problems, including congenital defects in the formation of the heart and valves.

Turner syndrome is a particularly good example of the limitations of "intersex" as a catch-all term for DSDs. Women with Turner syndrome are not sexually ambiguous from a physical perspective—although this condition is certainly a disorder of sexual development in that it impedes the development of a fully functioning female reproductive system (they can have sexual intercourse like other women) and secondary sexual characteristics—nevertheless, without medical intervention, they are sterile.

Congenital adrenal hyperplasia, or CAH, is a disorder that results in faulty regulation of hormone production. In its most severe form, it can also cause devastating dehydration, due to salt wasting, in male and female infants alike that, when undetected, can lead to death within a few weeks of birth. Typically, male infants with CAH are at greater risk for this condition because they are born with normal-appearing genitalia so CAH may not be diagnosed before a dangerous degree of dehydration sets in. In its milder forms, especially in genetic females (XX karyotype), it can cause ambiguous genitals. In those cases, an overproduction of androgens causes them to develop some external male sexual characteristics. The physical manifestations of CAH range significantly, depending on its severity. This can range from a slightly larger than normal clitoris and a slightly smaller than normal vaginal opening to what appears to be a penis and scrotum. In the latter case, the scrotum is empty, because there are no testicles. Increased androgens during fetal development has steered the direction of external physical development into somewhat of a male direction, but without the presence of the SRY gene (from the Y chromosome), the gonads themselves head down the ovarian route of development. At the same time, the high levels of androgens prevent the female urethra from forming normally, causing some of these girls to urinate through their clitorises. In these instances, because of physical appearances, the child is occasionally, at least initially, thought of as male and assumed to suffer from undescended testicles. Some forms of CAH also require surgery to close the opening between the uterus and bladder. If it's needed, a vaginal opening is often created at the same time.

Other instances of CAH are milder and do not even manifest until later in childhood. Baby girls may appear normal at birth, but as they grow older and androgen levels rise, they begin to display some male sexual characteristics. They may develop increased muscle mass, facial hair, and even enlarged clitorises.

A condition that occurs in genetic males and can result in ambiguous genitalia is called 5-alpha reductase deficiency. It is unusually common in some communities, including one in the Dominican Republic, where one mother of ten reportedly has four children with the condition. Some forms of 5-alpha reductase deficiency manifest rather dramatically at puberty, when children who were born appearing to be girls start to develop male sexual characteristics, including a penis, and testicles that suddenly descend. The villagers actually call the condition *guevedoche*, or "balls at twelve."

As you're no doubt beginning to realize, there is a broad gray area between indisputably male and indisputably female. Any combination of male or female traits from chromosome pattern to genitals to reproductive systems to secondary sexual characteristics can produce an individual who is not strictly male or female. And on top of all of that, there are those people who seem to be born entirely male or female but who feel anything but.

The magazine *New Scientist* tells the gripping story of a woman they call Sally Jones and her son Nick (their names were changed to protect their privacy). Nick has insisted he is really a girl since he could first speak; when he was five he announced that "God has made a mistake." He regularly dressed as a girl and began scratching his skin raw when he got an erection. In 2006, when he was thirteen, Sally found him holding a knife getting ready to amputate his penis.

Nick was diagnosed with gender identity disorder (GID), a rare but terribly traumatic condition in which there is a complete disconnect between one's physical sex and one's self-perceived gender. This condition is diagnosed when there are no physical or genetic gender inconsistencies, when the disconnect is between body and mind. Gender identity disorder is involved in an individual's sense of core identity. It is part of the answer to the question "Who am I?"

Alison George, a biologist and editor at *New Scientist*, writes:

So what is GID? It is not simply a case of boys being effeminate or girls being tomboys—although affected children do reject the toys, activities and clothing typical of their gender.

Boys with GID often assert that their penis is disgusting and will disappear. Girls commonly claim that they will grow a penis and say that they do not want to develop breasts or menstruate. Essentially, gender identity is how someone perceives and identifies themselves that is irrespective of both biological sex and sex orientation.

Exactly how gender identity occurs is a matter of much debate and disagreement. Some believe that people whose gender identity is at odds with their physical bodies and who apparently do not have an underlying genetic or hormonal disorder have a purely psychological condition. But others believe that there may be other physiological influences. A new study, published in 2008, by Austrian scientists found that specific versions of the gene, CYP17, were found to be more common in female-to-male transsexuals, than in women who did not identify as transsexuals. It's thought that female-to-male transsexuals may have higher levels of testosterone because of the version of CYP17 they have.

With a lot of help from advocacy groups like ISNA, and as new research embraces the possibility that gender identity may be physiologically rooted in the brain and not just the genitals, the medical community has begun to adjust its approach. The old approach might best be characterized as "cut first." When babies were born with ambiguous genitals, doctors and parents would "assign" a sex to a given infant, perform cosmetic surgery to bring the infant's genitals into "conformity" with its new assignment, and hope for the best. Of course, if you believe that gender identity is the result of a complex neurological stew in which genetics and hormones and environment all play a part, it's not hard to imagine those sex reassignments going awry fairly often. Which they have.

There's the story of a seven-month-old Canadian boy named Bruce whose routine circumcision went so badly awry that doctors recommended sex reassignment as a girl. His parents took their advice, Bruce underwent further surgery to fashion female genitals, and his name was changed to Brenda. By the time he was fourteen, he had completely rejected feminine clothes and behaviors and called himself David.

Dr. Eric Vilain, professor of urology and human genetics at UCLA, is a member of the ISNA medical advisory board and one of the leading researchers into the genetics of sexual determination and identity. He is also on the forefront of those fighting to make sure that individuals with these conditions are treated with a level of thoughtfulness and concern for a patient's lifelong well-being and happiness. This approach goes beyond the hush and rush philosophy of the past—keep the disorder quiet and rush quickly to a sex reassignment surgery. Vilain is currently studying how gender identity may be chemically established in the brain through mechanisms that are distinct from the sex hormones and gonads. "What really matters is what people feel they are in terms of gender, not what their family or doctors think they should be," says Vilain.

The work of groups like ISNA and scientists like Vilain has led to an emerging move to hold off on appearance-related surgery until the child is old enough and mature enough to make decisions on their own, with the advice and support of family and professionals.

If gender identity is at least partially rooted in the brain, what about sexual orientation? And if sexual orientation has some genetic roots, how has it persisted in the gene pool, given evolution's preoccupation with reproduction and the obvious challenges homosexuality brings to that endeavor?

let it be

n 1972, Linda Wolfe, an aspiring graduate student studying anthropology at the University of Oregon, was in need of the perfect dissertation topic. A psychologist she happened to know,
who had spent some time studying a troop of Japanese macaques,
told her that the macaques were spending a great deal of time
sexually pleasuring each other. Well, big deal. Besides eating and
sleeping, what else are monkeys supposed to do but make babies,
right? Right—except the sexual contact this psychologist described
had nothing to do with making babies. It was female-to-female.
Lesbianism ruled the troop.

Now, the macaque troop in question had a shortage of males,
so researchers at the time thought the lesbian behavior might be
analogous to same-sex relations that occur in prisons. Given only
one outlet for sexual contact, many choose same-sex over nothing at

all. So Wolfe took the obvious next step: she went off to Japan to study a troop of macaques with a more balanced ratio of males to females. You guessed it. Again, lesbianism ruled the troop.

That's not all Wolfe discovered. Macaques were very into female-to-female sexual contact, but they weren't indiscriminate about their sexual partners. Like humans, they were choosey. And, like humans, they avoided sexual contact with close relatives. She observed sexual contact between cousins, but never between mothers and daughters or grandmothers and granddaughters. One popular theory regarding homosexual contact among animals (yes, it's not just macaques, as we'll see shortly) is that one individual asserts dominance over another by forcing the contact. But that wasn't the case here; when the female macaques paired off, it was a two-way street; they seemed to pleasure each other.

Evolutionary biologist Paul Vasey is a leading homosexuality researcher. Like Wolfe, he has spent many years studying lesbian behavior in Japanese macaques. Female homosexual sex in macaques isn't just random; macaques engage in elaborate courtships that can last up to a week, although they also engage in the monkey equivalent of a quick afternoon romp.

So what's going on with these macaques? Japanese macaques live in a matriarchal society; females are dominant. Some researchers believe that the rampant female-to-female sexual contact among macaques helps to reduce aggression and provide social bonding. It obviously doesn't foster reproduction per se—something else is happening. As Vasey succinctly states: "Traditional evolutionary theories for sexual behavior are inadequate and impoverished to account for what is going on."

Linda Wolfe thinks the reason behind all the lesbian sexual contact is straightforward—they do it because they like it. Today, Professor Wolfe is chair of the Department of Anthropology at East Carolina University. When I asked her what she thought about the evolutionary origins of homosexuality, she said, "It probably has no evolutionary significance. The females still get pregnant no matter if they engage in occasional homosexual activity, and even continue to do so [engage in homosexual activity] after they're pregnant as well." I asked if her years of studying sexuality in Japanese macaques had shed any light on human sexuality. "It's just part of the repertoire of what primates do," she said. "When it comes to macaques, who knows what's inside their heads? We can't ask them, but it may just be part of the way they enjoy to pleasure themselves and each other."

Macacques are certainly not alone in terms of homosexual contact in the animal kingdom. In fact, as Joan Roughgarden, professor of biological sciences and geophysics at Stanford University, documents in her book *Evolution's Rainbow*, homosexual contact is pretty rampant among all sorts of animals. Roughgarden has a different take on evolution and homosexual behavior than Wolfe. She thinks homosexuality has an evolutionary purpose, just not one that's related to reproduction. Instead of asking how homosexuality could persist when evolution is only concerned with reproduction, she points to the vast incidence of homosexual acts among animals and thinks we should be asking whether sexual contact serves some other evolutionary purpose besides reproduction:

> My discipline teaches that homosexuality is some sort
> of anomaly. But if the purpose of sexual contact is just

reproduction, as Darwin believed, then why do all these gay people exist? A lot of biologists assume that they are somehow defective, that some developmental error or environmental influence has misdirected their sexual orientation. If so, gay and lesbian people are a mistake that should have been corrected a long time ago. But this hasn't happened. That's when I had my epiphany. When scientific theory says something's wrong with so many people, perhaps the theory is wrong, not the people.

Roughgarden believes that homosexuality is actually a trait that evolution has preserved in species after species because, like heterosexual contact, it promotes intimacy, fosters bonding, and defuses aggression and tension, all of which increase the likelihood of survival for members of a species. Her book examines homosexual contact in many species to press her case. Here's some of what she describes:

> Male big horn sheep live in what are often called "homosexual societies." They bond through genital licking and anal intercourse, which often ends in ejaculation. If a male sheep chooses to not have gay sex, it becomes a social outcast. . . .
> Giraffes have all-male orgies. So do bottlenose dolphins, killer whales, gray whales, and West Indian manatees. . . .
> Bonobos, one of our closest primate relatives, are similar [to macaques], except that their lesbian sexual encounters occur every two hours. Male bonobos engage in "penis fencing," which leads, surprisingly enough, to ejaculation. They also give each other genital massages.

Roughgarden worked with a curator and artist to create an

exhibition entitled "The Sex Lives of Animals" at the Museum of Sex in New York City. The exhibition, which opened in July 2008, described animals masturbating, performing oral and anal sex, and even acting like exhibitionists. It explores homosexual contact among lions, giraffes, elephants, bison, and dolphins. In short, it makes it incredibly clear just how enormous the sexual diversity of the animal kingdom really is.

All told, homosexual behavior has been documented in hundreds of mammals, birds, and other species. Cataloging animal homosexuality, biologist Bruce Bagemihl published *Biological Exuberance: Animal Homosexuality and Natural Diversity*, a 750-page encyclopedia that documents homosexual behavior in more than 470 different animal species. And the type of behavior runs the gamut: "Nearly every type of same-sex activity found among humans has its counterpart in the animal kingdom," writes Bagemihl. And, if you have any doubt about that, consider what Dutch researcher Kees Moeliker witnessed: homosexual duck necrophilia. That's right, Moeliker watched a male duck copulate with a dead male duck for an hour and fifteen minutes, and caught it all on film.

BONOBOS ARE A type of chimpanzee that live in central Africa. They are thought to be our closest living relative. And they have a more varied sex life than just about any other animal out there. "If you're looking for homosexual sex in vast quantities, forget humans," says British primate scholar and evolutionary biologist Robin Dunbar. "It's bonobos you want. It's scandalous. They'll have sex with anyone, never mind the sex or age."

Bonobos live and search for food in large groups of twenty or more. All that bumping and jostling as they compete for food and attention can lead to lots of tension, which is why many scientists who study them believe their constant sexual interaction may be designed to soothe anxieties and reduce those tensions. Dunbar goes on to say:

> One plausible explanation is that all this is principally a bonding device. . . . They'll be always bumping into one another, treading on each other's toes, and noticing that Jemima over there's got a temptingly nice fig; they need something that will diffuse conventional stresses and re-build relationships after squabbles. . . . Where we bring chocolates and flowers, they groom and kiss instead. . . . The idea is that the relaxing, rewarding qualities of sex have been captured for social purposes, to reduce conflict and hold the group together.

"Same-sex sexuality is just another way of maintaining physical intimacy," says Roughgarden. "It's like grooming, except we have lots of pleasure neurons in our genitals. When animals exhibit homosexual behavior, they are just using their genitals for a socially significant purpose."

Some of this thinking about same-sex sexual contact in animals shares common ground with theories about the evolutionary purpose behind female orgasm. That is, they both suggest that sex has a role facilitating bonding and positive social cohesion that goes beyond reproduction.

So if all this homosexual behavior is rampant among animals, why isn't it more widely known? "Although the first reports of homosexual behavior among primates were published more than 75 years ago, virtually every major introductory text in primatology fails to even mention its existence," says primatologist Paul Vasey. It may very well be that politics and personal predispositions have gotten in the way of reporting these behaviors. As biologist Valerius Geist admitted more than two decades ago, "I still cringe at the memory of seeing old D-ram mount S-ram repeatedly. To conceive of those magnificent beasts as 'queers.' Oh God! . . . Eventually I called the spade a spade and admitted that the rams lived in essentially a homosexual society."

WHEN IT COMES to humans, of course, the great debate over sexual orientation is one of origin. Nature—predetermined by genes or other biological factors? Or nurture—the product of an individual's environment and its subsequent influences? A combination of both—or simply a choice? Few in the scientific community believe it to be purely a choice or purely biological. The scientific literature abounds with significant correlations between possible biological factors and sexual orientation of individuals—but it's also filled with examples of people whose orientation contradicts those expectations. Remember, correlation means researchers have identified a pattern indicating that one or more traits tend to be associated; it does *not* mean that those traits *can* influence one another.

One of the most prominent—and controversial—studies to suggest a biological correlation to homosexuality is a study of male twins authored by two psychologists, Michael Bailey of Northwestern University and Richard Pillard of Boston University. According to their study, if a male identical twin was gay, the odds of his brother being gay were around 50 percent—as much as ten to twenty times more likely than someone in the general population. Even in pairs of male fraternal twins, where siblings only share about half their genes with each other, the psychologists found that if one brother was gay, the odds of the other brother being gay were about 20 percent. Of course, just as this study suggests a biological correlation, it also seems to rule out genetics as the *only* factor in determining orientation.

These studies aren't the only ones that suggest a biological source for sexual orientation; there have been many others, like the scent studies discussed earlier, showing how the brains of homosexual men process male scents the same way that heterosexual women do, and the brains of lesbian women process female scents more like heterosexual men.

Many of these studies are underfunded. This means they are often relatively small, which, in turn, makes them less conclusive. The smaller a study is, usually the less statistically certain its results. But many scientists believe that, taken together, all these studies form a composite picture that looks a lot like a biological signpost, albeit a small one. The research "all sort of pointing in the same direction makes it pretty clear there are biological processes significantly influencing sexual orientation," says neuroscientist and author Simon LeVay. "But it's also kind of frustrating that it's still a bunch of hints, that nothing is really as crystal clear as you would like."

If these researchers are right, and there *is* a biological link to homosexuality, besides assisting with social cohesion, how did it survive in the gene pool? In other words, if there's a gay gene or genes that predispose an individual to nonreproductive homosexual behaviors, how does that gene continue to get passed on instead of dying out?

An Italian study offers an intriguing possibility: What if a genetic combination that makes a woman more fertile, or fecund as researchers call it, also works to incline a man toward homosexuality? Professor Andrea Camperio-Ciani and a team of researchers at the University of Padua interviewed ninety-eight gay and one hundred straight men, getting detailed family histories from them covering some 4,600 relatives. And they found that the female relatives of gay men, mothers and maternal side aunts, produced significantly more children than the mothers and maternal side aunts of straight men. That the difference was found only on the maternal side suggests that the genetic link, if there is one, may be found on the X chromosome, which males only receive from their mothers. Dr. Camperio-Ciani's latest 2008 study found the same correlation in the families of bisexual men—their female relatives had more children. Again, it's important to note that just because the researchers found a *correlation* in these studies, it doesn't actually mean they found a genetic *connection*.

Still, Dr. Camperio-Ciani is downright bullish about his work, feeling that it sheds light on the evolutionary mystery of homosexuality: "We have finally solved this paradox," says Camperio-Ciani. "The same factor that influences sexual orientation in males promotes higher fecundity in females."

The possibility certainly merits further investigation and has a logical appeal from an evolutionary perspective. It's a lot easier to imagine the long-term survival of a gene promoting female reproduction that sometimes gets in the way of male reproduction than it is to imagine one that simply acts against male reproduction all the time. Commenting on the first Italian study, Dr. LeVay said: "We think of it as genes for 'male homosexuality,' but it might really be genes for sexual attraction to men. These could predispose men towards homosexuality and women towards 'hyper-heterosexuality,' causing women to have more sex with men and thus have more offspring."

"It helps to answer a perplexing question—how can there be 'gay genes' given that gay sex doesn't lead to procreation?" says Dean Hamer of the National Institutes of Health. "The answer is remarkably simple: the same gene that causes men to like men also causes women to like men, and as a result to have more children."

Even if it's right on the mark, the Italian study only tells a part of the story. Dr. Camperio-Ciani estimates that the effect his report describes accounts for less than a seventh of male homosexuals: "Our findings are only one piece in a much larger puzzle on the nature of human sexuality." Nevertheless, it's intriguing—evolution selecting for a gene that encourages women to have more children, even if some of them will be males with some inclination toward an orientation that takes them out of the reproductive pool.

As our understanding of just how genes work has deepened, scientists have started looking for clues to that larger puzzle beyond just gay genes. "Genetics is not determining the sexual orientation, it's only influencing it," says Camperio-Ciani. And recently, the search for a biological component to sexual orientation has focused on the influence of a baby's fetal environment—that is, the womb—on its genes and development. For most of its history, the field of genetics has believed that genes were somewhat immutable blueprints, directing your body to build itself according to set plans. In the last few years, though, our understanding of the interaction between genes and the environment has completely changed. We now know that certain genes can actually be turned on or off, much like flicking a light switch. One of the ways that this can occur is through a chemical process called *methylation*. All kinds of things can have an influence on this process, from cigarettes, to diets low in critical nutrients, as well as diets rich in others. And the process is clearly at work during pregnancy—the environment a mother lives in, her diet, even her anxiety can turn specific genes on or off in her developing fetus. Certainly this process could have an effect on genes that influence sexuality, as various factors influence the genes turning on or off.

As discussed in Chapter 5, one of the ways the expression of a given genetic sequence can be implemented or suppressed is through hormones and the body's receptivity to them. Women with androgen insensitivity syndrome, for example, have the genetic code to be men—they are XY—but, because they lack the hormonal receptors necessary to respond, they don't react to the testosterone that would otherwise trigger masculinization in the womb, and they follow a largely female developmental path. So although they are genetic males, the majority of women with AIS (XY), like genetic females (XX), are generally thought to have heterosexual preferences, and are thus attracted to men.

University of Oklahoma urologist Dr. William Reiner has examined more than a hundred cases in which genetic males—XY—were born with very underdeveloped or missing penises, subjected to surgical castration and genital "reconstruction" as females, and raised as girls. "I haven't found one who is sexually attracted to males," says Reiner.

This leads some researchers to believe that sexual identity and maybe sexual orientation are established in the womb before the genitals are even formed, as the brain is washed in waves of sex hormones. Reiner elaborates, "Exposure to male hormones in utero dramatically raises the chances of being sexually attracted to females. We can infer that the absence of male hormone exposure may have something to do with attraction to males."

More evidence that prenatal hormone exposure has some connection to sexual orientation is in your hands. Literally. Generally speaking, men have index fingers that are slightly shorter than their ring fingers. For women, the two fingers are usually closer in length. Scientists from the University of California at Berkeley have shown that the fingers of lesbian women follow the same pattern as men, with shorter index fingers. They think this may be the result of exposure to higher than normal levels of androgens, male sex hormones, during fetal development.

Seems like it's starting to make sense, right? Genetic men (XY) who don't respond to testosterone develop physically as females and are attracted to men. Females who are exposed to high levels of male hormones have male-pattern fingers and are attracted to other females. Not all men who are exposed to especially high levels of male hormones have especially male-pattern finger lengths, which follows the pattern—but they also tend to be attracted to other men, which doesn't.

That's right, some researchers think that men who are exposed to especially high levels of male hormones tend to be attracted to other men. It's actually not quite that simple—but it's more evidence that the chemical stew that influences sexual orientation, if it does, involves a pretty complex set of ingredients. The same study of finger length showed that gay men with several older brothers tended to have much shorter than normal index fingers relative to their ring fingers—in other words, the same phenomenon that masculinized the fingers of lesbians seemed to be at work but only if they had two or more older brothers. What's particularly interesting about this correlation is that previous research has shown that having several older brothers actually increases the odds that the subsequent child will be gay. The researchers believe that somehow there is an increase in the amount of androgens a mother's subsequent baby is exposed to, resulting in the "extreme" male finger patterns, and perhaps influencing orientation as well. Androgens, remember, are the hormones that can trigger male development. How exactly that happens, of course, is still very unclear.

But the one thing almost everyone agrees on is that sexual reproduction is incredibly complicated and there are numerous ways for things to turn out—from differentiation in the genitals, the rest of the body, or the brain; in the chemistry of identity, attraction, and desire; and, of course, in the complicated biological merger of genes from two parents, starting the whole miraculous process over again.

Which brings us full circle again—why sex?

tainted love

n 1995, when I was a college student, I spent the summer working at an orphanage in Bangkok, Thailand, called Tarn Nam Jai. It was an HIV/AIDS orphanage. All of the children there had lost their mothers to the epidemic that has ravaged so much of Africa and Asia. And the orphanage wasn't just an orphanage; it was a hospice too. About a quarter of children born to HIV-positive mothers become infected with the virus themselves. So Tarn Nam Jai had a double challenge—caring for the children with HIV and finding homes for the children who weren't infected. Of course, the very fact that Tarn Nam Jai had two missions was a miracle in itself. How did three-quarters of the children, all born to women with HIV, escape infection?

The placenta-uterus interface is an extraordinary viral filter, and normally about three-quarters of the time is able to prevent the

HIV virus—an especially insidious string of "genetic code"—from passing from mother to child. The fact that pregnancy can prevent HIV from infecting a fetus is a powerful testament to its effectiveness, but that's just a part of the story. HIV is so insidious because of the way it replicates itself, or reproduces. It's a retrovirus; it uses an enzyme called *reverse transcriptase* to actually write itself into the DNA of the person it has infected, becoming a permanent part of his or her genetic code. So every time a person's infected cell reproduces, HIV reproduces along with it. Unlike sexual reproduction (which requires a male and a female), asexual reproduction occurs when an organism makes an exact copy of itself. If a human infected with HIV reproduced this way, it would copy HIV right along into its offspring. Every time.

Sexual reproduction essentially helps us to wipe the biological slate clean; it helps to prevent us from passing along whatever infections we may have acquired in our lives. And that's not all. Sexual reproduction allows for genetic reassortment, almost like shuffling a deck of cards but on the genetic level. Every time we shuffle the deck, there's some chance of a new combination, one that may produce a stronger trait in our offspring, like an immune system that can outwit a new virus or bacterium. Asexual reproduction doesn't usually allow for a reshuffling of the genetic cards, but more or less ensures that an organism's offspring will be no better off than its parent.

Sex gives the species two big advantages in the evolutionary sweepstakes: it can at times protect children from the responsibility for their parents' biological mistakes. It gives those children the chance to biologically outdo their parents. That's why sexual reproduction may be pretty rampant throughout nature. Unfortunately, it also allows a mother to harm her child, through alcohol and/or drug abuse, before the child is even born. Yes, there are all kinds of ways sex can go wrong, and yes, it comes with all kinds of costs. And yes, it's expensive.

But you get what you pay for.

SEX HELPS US to give our children a clean bill of health, working to spare them from a lifetime's accumulation of parasites, viruses, and other pathogens. But remember, evolution is all about trade-offs, and the world is filled with opportunistic creatures, some large and some very, very small. Another significant cost of sexual reproduction is that the very act of sex opens up a whole new niche for some of those creatures to move from one host to the next. File that under life's little ironies: sex protects our children from inheriting infections while exposing us to the risk of more infection in the process.

And, of course, it doesn't provide perfect protection, not by a long shot. As I said, the miracle I witnessed at Tarn Nam Jai was that three-quarters of the children escaped infection. But heartbreak was the flip side of the miracle—for every three children born healthy there was a child born dying. The face of one of them has never left me. His name was Johnny, he was five years old, and he had full-blown AIDS. His body was unbelievably frail, his belly was grossly distended, but most of all, he was in so much pain, dying shortly after I left Thailand.

Watching children die of HIV left me with a clear conviction: in order to stay one step ahead in the evolutionary arms race, nature may have led us down a path filled with perils, but that doesn't mean we can't defend ourselves against some of them, and sexually transmitted infections is a good place to start.

Sexually transmitted infections can range from the mildly annoying, like the pubic louse, to the seriously life-threatening, like HIV.

STIs have typically been associated with a stigma not attached to many other diseases, which makes it all the more important to have a clear understanding of how they're transmitted, the risks they pose, and what their warning signs are.

The most common sexually transmitted virus in the United States is actually one that many people are unfamiliar with, although it's gotten a fair amount of attention in recent years, not so much because of the damage it does on its own but because of the additional risks it poses for people infected with it, especially women. That virus is *human papilloma virus*, or HPV.

At least a quarter of American men and women have been infected with HPV. 6.2 million new infections happen every year. HPV can often infect without causing any symptoms, but in some cases it causes genital warts—warts with a cauliflower-like appearance, commonly found on the penis, vulva, and anus. Sometimes those warts can go away on their own, but if they don't, they can be removed by a doctor, much in the way warts on the foot or hand are removed, through surgery, cauterization, or cryotherapy (freezing them off with liquid nitrogen). So far, there's no complete cure. In severe cases HPV can cause genital warts that can be very disfiguring and quite traumatic. But it's not HPV's manifestation as warts that has earned it headlines recently.

HPV is the main cause of cervical cancer in women, the second leading cancer killer of women worldwide. In the United States, cervical cancer is diagnosed in about 10,000 women every year, and 3,700 die from it. HPV has also been linked to oral, rectal and anal cancers in both men and women. With the increase in prevalence of oral and anal sex amongst heterosexual couples, it's thought that we may see a corresponding increase in HPV-related cancers.

Earlier this year, I got to see first-hand the devastation that an untreated HPV infection can cause. A patient was admitted to a New York City hospital because of an abscess in his groin that was a little larger than a tennis ball. Unfortunately for him, the abscess was the least of his problems. He had a pretty advanced form of penile cancer that spread to his groin, causing the abscess. Most of the head of his penis (called the glans) had been overtaken by cauliflower-like warts. Much of his penis had to be removed through an amputation procedure called a *penectomy*. Unfortunately for this patient, although he was still young, he has a less than 50 percent chance to live another five years.

AS WE DISCUSSED in Chapter 2, HPV has also been at the center of one of the biggest breakthroughs in cancer treatment we've seen. Like most viruses, HPV comes in many different strains—more than one hundred have been identified so far. Last year, the Food and Drug Administration gave its approval to a vaccine that prevents infection from four of the most common strains—HPV types 6, 11, 16, and 18. Two of these, types 16 and 18, are thought to be responsible for around 70 percent of all cervical cancer, whereas types 6 and 11 cause and 90 percent of genital warts. The medical community now recommends vaccination against HPV for all girls starting at eleven or twelve years of age, although the vaccine can be administered to girls as young as nine. So far the FDA has only approved the vaccine for women up to age twenty-six, although some doctors have been giving it off-label to older women. The key to successful protection is vaccination before a girl's first sexual contact. And sexual contact means exactly that—contact, not intercourse.

Like many STIs, including herpes, gonorrhea, syphilis, chlamydia, HPV can also be transmitted through oral sex. In fact, oral cancer—cancer of the tongue, mouth, and throat—caused by HPV has climbed sharply over the last three decades, and scientists link the climb to increased oral sex, which means oral contact with a sexual partner's penis, vulva, vagina, or anus.

A team of researchers at Johns Hopkins University in Baltimore, Maryland, studied forty-six thousand cases of oral cancer from the last thirty years. They report that oral cancers linked to HPV climbed by more than 30 percent from 1973 to 2004, even as oral cancers unrelated to HPV have dropped since 1982, probably due to a drop in other risk factors like smoking, chewing tobacco and drinking alcohol, according to the researchers. The sharpest rise in oral cancer was among young white males. This makes sense to other experts, like Lesley Walker, the director of cancer information at Cancer Research UK: "What we do know is that the prevalence of HPV is high, particularly among young people," says Walker, "and this shouldn't be a surprise given that, since the sexual revolution, people have been having more sexual partners."

"We need to start having a discussion about those cancers other than cervical cancer that may be affected in a positive way by the vaccine," says Dr. Maura Gillison, who heads up the Johns Hopkins research team. She believes there's a strong argument to be made for vaccinating young boys as well as young girls against HPV.

Not everyone is so sure. Walker thinks more research is necessary to ensure that it would actually make a significant dent in the incidence of male oral cancer in order to justify the high cost of vaccinating so many boys. Still, there are many who argue it's a good idea, and not only because of oral cancer. Cervical cancer is so deadly to women that some believe it makes sense to prevent as much HPV infection as possible, and that means vaccinating boys too. Obviously, if a man has been vaccinated against HPV and can't become infected, then he can't infect other woman (or men, since there is an association between rectal cancer and HPV), whether or not they've been vaccinated.

THE MOST COMMONLY reported bacterial STI in the United States today is called *Chlamydia trachomatis*. Somewhere in the neighborhood of 1 million cases of chlamydia are logged every year. It's especially common in sexually active adolescent girls and African American women between the ages of eighteen and twenty-six. And a recent study suggests that sexually active college freshmen of both sexes may be especially at risk.

Chlamydia is a particularly sneaky infection because it very often doesn't cause any symptoms; about half of infected men and 80 percent of infected women are asymptomatic. The microbe itself is hard for our immune system to find and kill because it typically resides inside human cells, whereas our immune system has an easier time dealing with entities outside of our cells. When it does cause symptoms in both men and women, it makes it painful to urinate. For some men, there can be a small discharge or a swelling of the testicles as well. Making matters even trickier, until recently, the best way to diagnose chlamydia in men was by examining scrapings of epithelial cells from the man's urethra. This involved inserting a Q-Tip–like probe into a man's urethra through the head of his penis. As you can imagine, this made some men reluctant to be tested. Thankfully, there are newer and much more comfortable ways, such as a urine test, to detect chlamydia today.

The symptoms of an infection aren't the most significant threat posed by the disease—it's the silent damage to fertility that makes chlamydia such a problem. In women, chlamydia can cause tubal factor infertility, scarring the Fallopian tubes or even blocking them entirely, making it difficult or impossible for sperm and eggs to meet. For a long time, doctors and scientists thought chlamydia only posed a threat to female fertility, but new research shows there's a real risk to men too. Two studies have found a significant link between chlamydia and male infertility.

In 2004, a team of scientists led by Professor Jan Olofsson of Umeå University Hospital in Sweden published a report in the journal *Human Reproduction* that found a correlation (remember, this does not mean causation) between fertility trouble and chlamydia infection in men. Olofsson and his team tracked 238 couples getting help for fertility problems from 1997 to 2001. They found that men infected with chlamydia were about one-third less likely to become fathers than men who were not infected.

In 2007 José Luis Fernández of the Juan Canalejo University Hospital in Corunna, Spain, and his colleagues presented a report to the American Society for Reproductive Medicine that helped explain the connection between chlamydia and male infertility. Fernández and his team found that chlamydia seems to cause severe genetic damage to the sperm of infected men. When they compared the genetic health of sperm in infected and uninfected men, they found that 35 percent of the sperm from infected men had fragmented DNA, compared to only 11 percent in the uninfected men. The good news is that treatment with antibiotics not only cured the men of chlamydia; it helped to restore their sperm to a healthy state.

Besides infecting the reproductive system, chlamydia loves to get into people's eyes. Once there it causes conjunctivitis, which in its chronic form becomes trachoma, probably the leading cause of blindness in the world. In some parts of the world, these types of chlamydial infections are actually spread by flies. But the surest way for children to become infected is from their mothers.

HERPES SIMPLEX IS actually two related DNA-type viruses—herpes simplex virus 1, or HSV-1, and herpes simplex virus 2, or HSV-2. Herpes infections were once segregated by the physical site of infection. The two most common types being oral and genital herpes.

HSV-1 is commonly thought to be the virus that causes oral herpes and HSV-2 the virus that causes genital herpes, but that's not totally correct. As a real estate agent might say, it's location, location, location. Either virus can infect above the waist or below it. In fact, HSV-1 infections of the genitals have been steadily rising, perhaps in part because of increased rates of oral sex.

Herpes is a neurotrophic virus; it loves to make its home in the nervous system. Interestingly, it's thought that humans are the only animals "hospitable" enough to give herpes a home.

In the United States, the number of people with antibodies to HSV-1 is as high as 80 percent. When the immune system encounters an infecting agent like a virus, it produces specific proteins, called antibodies, that help fight the specific infection. Hence, the presence of antibodies indicates past exposure. Prevalence increases with age across the board, but there are differences among socioeconomic groups. By age thirty, for example, half of the people in upper socioeconomic groups have antibodies to HSV-1, compared to 80 percent of people in lower socioeconomic groups. Lower socioeconomic groups tend to bear a higher load of diseases in general, partly, it is thought, because of less access to proper diet and health care.

A little over 1 million new cases of genital herpes are diagnosed every year, and more than 50 million people—around one in four adults—are thought to be currently infected in the U.S. If that doesn't seem to add up, it's because herpes never goes away. Once people are infected with herpes, they are thought to be infected for life, although they don't always have symptoms and they're not always contagious.

Both oral and genital herpes (regardless of the subtype of virus that caused them) manifest symptomatically as lesions or blisters filled with the virus. Oral herpes appears on the lips, tongue, cheeks, and gums. Sores from genital herpes can break out on the penis, the vulva, the anus, and even the inner thighs and buttocks. In healthy people, the sores usually resolve and heal within a few weeks. At this point the virus migrates back into nearby nerve tissue, where it resides without causing symptoms. Herpes cycles between two stages: the latent stage, or remission, and the symptomatic stage, known as active disease. This period of latency is one of the many evolutionary strategies that some microbes employ to escape detection by our immune systems and wait for opportunities to infect others.

During phases of active disease, herpes is highly contagious, and contact with infected sores should absolutely be avoided. One of the tricky things about genital herpes, though, is that it can also be contagious even when the person is asymptomatic. The virus can still shed infectious copies of itself, even without causing symptoms. Some people, probably because of genetic background or strong immune responses (possibly a combination of both) can be infected with herpes and never actively express any symptoms, although, as just mentioned, they can still be contagious. Others have multiple recurrences per year. People infected with either subtype of herpes, and their partners, should use condoms and dental dams (a piece of latex you place over the vulva or anus) to prevent transmitting the virus. We don't know exactly what causes recurrences, but stress, exposure to sunlight, having another (different) infection, and even menstruation have all been identified as potentiating factors. It's almost as if the virus waits for the most opportune time when the immune system is somewhat compromised to mount an attack.

Herpes doesn't infect just the oral or genital regions; it comes in other varieties too. *Herpes whitlow* is a very painful infection of the fingers or toes that is most often contracted by health-care workers exposed to the virus, especially by dental workers who encounter oral herpes. It is characterized by the emergence of small, blisterlike sores that merge as they swell and cloud. The sores typically last two to three weeks. Herpes can also infect the eye; it's then called ocular herpes or *herpes keratitis*. There's even a version called herpes gladiatorum, *wrestler's herpes*, or mat herpes, which affects people in contact sports. When scrapes, cuts, or mat burns on the skin come into contact with the herpes virus, an individual can become infected and suffer sores at the site of infection.

And herpes simplex encephalitis, or HSE, is an infection of the brain and nervous system. Exactly how herpes travels through the nervous system to gain access to the brain isn't fully understood, but the effects and danger posed by HSE are well known. Untreated, HSE is fatal in more than two-thirds of cases; even treated, it kills about one in five, and it can leave those infected with some form of brain damage.

One of the very rare effects of herpes encephalitis is at once disturbing and intriguing. Damage from herpes (or other trauma) to a part of the brain called the amygdala, which is thought to be responsible for emotional learning, can produce a rare, yet sometimes reversible, neurological disorder called Klüver-Bucy syndrome.

People with Klüver-Bucy syndrome routinely explore objects with their mouths (which is a great way for the virus to spread), much like an infant or toddler. And they often become hypersexual, displaying very inappropriate sexual behavior. About a third of HSE infections are in children under the age of eighteen. When infection results in Klüver-Bucy syndrome, the symptoms of hypersexuality can be troubling to observe.

I believe the virus's attack on the amygdala is no evolutionary accident. As I wrote about at length in *Survival of the Sickest*, there is clear evidence throughout nature of pathogens affecting the behavior of their hosts in a way that facilitates their transmission. In the rare case of Klüver-Bucy, the hypersexuality that can result may be quite extreme.

New York psychiatrist Laurence Tancredi explains that the behavior patterns we consider aberrant may sometimes have biological drivers. "In children, [Klüver-Bucy syndrome] may be manifested by intermittent thrusting of the pelvis, holding of one's genitals, or rubbing the genitals in a masturbatory movement on the bed," writes Tancredi in his book *Hardwired Behavior*. What better way for a sexually transmitted infection to improve its chances of infecting new hosts than to push its host toward promiscuity by triggering hypersexuality?

Klüver-Bucy syndrome is exceptionally rare, and, thankfully, these behaviors seem to subside on their own over time. What is interesting and still unclear is whether the virus that causes herpes can cause subtle behavior changes without causing full-blown encephalitis.

The truth is, in most people with healthy immune systems, herpes can be terribly unpleasant, but it's very rarely life-threatening. But it's very important to be aware if you're infected, especially if you're pregnant. Herpes can be transmitted to newborns during delivery, which is thought to happen as frequently as once in every 2,500 live births in the United States. In newborns, a herpes infection is very dangerous. Death from neonatal herpes can be as high as 60 percent of those infected. If a woman has a primary infection of genital herpes—that is, she is symptomatic or asymptomatic but still shedding virus at the time of her initial infection—there is a 50 percent chance that her new baby will contract herpes while he or she is being delivered, if the delivery is vaginal. The risk can be reduced by a cesarean delivery, which is why it is so important to tell your doctor if you're pregnant and think you might have herpes. If you're having sex with a pregnant woman and think you might have herpes, tell your partner. The key thing here is to remember that symptoms don't matter—if you've *ever* had herpes, you always will have herpes and need to use that information to protect yourself and others, especially when it comes to the future health of newborns.

There's no real cure for herpes so far, although there are a few antiviral medications on the market that seem to shorten the length of outbreaks of active disease. If you're one of those unlucky people who have multiple recurrences per year, you should know that there are also antiviral regimens that can help reduce recurrences.

There may be other options too. A few studies have shown that increasing the ratio of lysine to arginine, both amino acids that you consume (either through the diet or with supplements), may help to reduce the severity and length of outbreaks. Lysine, in higher relative concentrations to arginine is found in certain foods such as red meats and dairy products; higher levels of arginine is found in wheat, many kinds of nuts, and fruits. The easiest way to increase the ratio of lysine to arginine may be to take lysine supplements, but you should talk to your doctor before doing so. Some animal studies have shown that lysine may increase your cholesterol and triglycerides (high levels of which, like cholesterol, have been shown to contribute to arteriosclerosis, the hardening of the arteries that can cause heart disease).

Even more promising, there is hope of a herpes vaccine on the research horizon. Studies have shown that an experimental vaccine provided about 73 percent protection in women who had never been infected before. That's certainly not perfect—and it doesn't work at all in men—but it's a start.

GONORRHEA, CAUSED BY a bacterium called *Neisseria gonorrhoeae*, is one STI that has actually been declining over the last twenty years, at least in the United States. But for anyone who does get infected, it's a serious concern. The highest rates of new infections are in young women between the ages of fifteen and nineteen.

Gonorrhea is another STI that can often remain hidden as its symptoms can be hard to spot or may not manifest at all. In men, the acute infection tends to become symptomatic two to seven days after infection. Symptoms can include pain during urination and discharge of a puslike substance sometimes called *gleet*. Less than half of women experience symptoms; when they do, there's often discharge. Even with discharge, these symptoms are often missed or confused with yeast infections. And, like other stealthy STIs, untreated gonorrhea can lead to some serious complications, including pelvic inflammatory disease in some 10 to 20 percent of cases, and also cause infertility in women. Gonorrhea can also lead to septic arthritis, which occurs when bacteria migrate to joints and cause painful swelling and damage. Gonorrhea can also be spread through oral sex, infecting the throat, called *pharyngeal gonorrhea*, and anal sex, called *rectal gonorrhea*.

Like other bacterial infections, gonorrhea is treated with antibiotics. Since co-infection with chlamydia is so common, antibiotics are usually given to treat both at the same time. And, like other bacteria commonly treated with antibiotics, gonorrhea has evolved antibiotic-resistant strains. The common antibiotic tetracycline is now essentially ineffective against gonorrhea, and ciprofloxacin-resistant strains are now common as well.

The slang name for gonorrhea is "the clap." There are a few theories about the origin of the term. The first, and most unpleasant to imagine, is that someone would "clap" the penis on both sides to clear the urethra of pus. Its origin may stem from the obsolete French word *clapoir*, which means "bubo," a swollen lymph node as one might experience from infection with gonorrhea (or the bubonic plague). It may also be related to another old French word, *clapier*, which means brothel. Or maybe it was just a combination of the two; people called it the clap because they had a *clapoir* associated with too many visits to the *clapier*.

PERHAPS THE STEALTHIEST of the most common STIs, syphilis has been known by many names, including a sort of nickname coined by the British medical scholar Sir Jonathan Hutchinson in 1879. He called it "The Great Imitator" because its symptoms can seem to mimic so many other diseases. That is, if you experience any symptoms at all.

Syphilis is also probably the only STI to have been immortalized in a three-volume epic poem. The Italian poet Girolamo Fracastoro actually gave syphilis its name in his 1530 Latin poem "Syphilis sive morbus gallicus" ("Syphilis, or the French disease"). The French disease? That's what the Italians and Germans called it, Fracastoro explains. And the French called it the Italian disease, the Dutch called it the Spanish disease, and Turks and Arabs called it the Christian disease.

Until Fracastoro named it syphilis, it was also commonly known as the great pox (as opposed to smallpox). A description of a syphilis sore written by Ulrich von Hutten in 1519 certainly makes the poxes sound large:

> Boils that stood out like Acorns, from whence issued such
> filthy stinking Matter, that whosoever came within the

Scent, believed himself infected. The Colour of these was of a dark Green and the very Aspect as shocking as the pain itself, which yet was as if the Sick had laid upon a fire.

Now, that sure doesn't sound like a stealthy disease, does it? And the modern experience of syphilis isn't like that at all. Today, when someone is first infected with syphilis, it usually produces a single blister, called a chancre, at the site of the initial infection. Syphylitic chancres are usually small, round, and painless; last for a few weeks; and then disappear without treatment—which many people don't seek, precisely because chancres are small and painless.

Chancres are symptoms of the primary stage of syphilis. Untreated, the infection moves into its secondary stage, usually six to eight weeks after infection. Symptoms during the secondary stage include a rash or rashes on the body. Typically they occur as a rough rash on the palms or the soles of the feet, although outbreaks can occur just about anywhere on the body and, in keeping with syphilis's status as a Great Imitator, often resemble rashes caused by other conditions. Like chancres, secondary stage syphilis will also resolve without treatment in most people with a competent immune system.

After primary and secondary stage symptoms clear up, syphilis moves into its latent stage where it can remain hidden for years. For many people, latent syphilis will never cause additional problems, but in about 15 percent of those infected with syphilis who do not receive treatment, syphilis can enter its very serious and sometimes even deadly third stage. Tertiary stage syphilis, which can appear decades after the initial infection, is characterized by the growth of tumorlike masses called *gummas* that often appear in the liver, but can manifest in the brain, heart, bone, and elsewhere throughout the body. Tertiary syphilis can cause dementia, paralysis, blindness, and death.

Medical historian Deborah Hayden sees a pattern of syphilislike symptoms in the lives of a diverse group of historic figures, including Adolf Hitler, Friedrich Nietzsche, Oscar Wilde, Ludwig van Beethoven, and Vincent Van Gogh. In her book *Pox: Genius, Madness and the Mysteries of Syphilis*, Hayden describes how these and nine other figures experienced a series of symptoms, from muscle and joint pain to internal disorders that were capped by varying degrees of late-life mental imbalance that she finds characteristic of syphilis.

But how did a disease that was described as one producing sores the "size of Acorns" end up as such a silent threat? Biologist Robert Knell of Queen Mary University in London believes that evolution favored strains of syphilis that didn't cause such visible sores. Knell argues that nobody would want to have sex with someone displaying large, pus-ridden sores. And since the bacterium that causes syphilis depends on sex for transmission, strains of the organism that left infected people in a more attractive state were likely to do better from an evolutionary perspective. If people with strains of syphilis that didn't cause big sores had more sex (because their prospective partners wouldn't detect the infection), those specific strains of syphilis could then become dominant and thus more common, and end up infecting more people.

Which just might have turned *Treponema pallidum pallidum*, the bacterium that causes syphilis, into the sneaky infectious player it is today. And it is yet another reason why you should talk to your doctor if you ever experience an unexplained sore on your genitals. Syphilis can be treated with antibiotics—but only if you know you've been infected.

LAST, BUT BY no means least, let's return to the STI that I saw infect too many young children. Human immunodeficiency virus, or HIV, the virus that causes acquired immune deficiency syndrome, or AIDS. Much has been reported about HIV and AIDS over the last twenty years, but a quick review of the numbers is in order. In the *2008 Report on the Global Aids Epidemic*, the Joint United Nations Task Force on HIV/AIDS and the World Health Organization report that 33 million people, including more than 2 million children, were living with HIV in 2007. More than 25 million people have died of AIDS since 1981, when the disease was first recognized. And there are 11.6 million AIDS orphans in Africa.

HIV, like so many other STIs, is also a sneaky stowaway, which helps it to move from one person to another. The first symptoms of HIV infection, which usually appear around two to four weeks after infection, look and feel like the flu, making diagnosis very difficult, even if the infected individual visits a health-care professional. HIV then moves into a latent period. Behind the scenes, though, HIV is very busy, winding its way into its host's DNA and taking up residence in the cells of the immune system. During this period of latency, which can last as long as twenty years or even longer, sexual activity gives the virus many opportunities to spread itself to new hosts—there are no boils standing out like acorns or purulent discharge to warn potential sexual partners that sexual activity poses a risk to their health. Eventually, most people infected with HIV develop AIDS, the complex of symptoms and infections that results from immune system failure caused by HIV infection.

Most people in the developed world understand that unprotected sexual intercourse is a risk factor for HIV transmission. But, surprisingly, there are still many people who don't recognize that this means lots of different kinds of sexual practices, including of course anal sex. A sweeping survey of girls of high school age by Dr. Avril Melissa Houston, who is now the deputy chief medical officer of Baltimore, revealed that 20 percent didn't know that penile-anal intercourse exposed them to increased HIV risk. In fact, a 2005 report by the Centers for Disease Control and Prevention (CDC) estimates that receptive anal sex is five times more likely to result in HIV transmission than traditional penile-vaginal intercourse. What about the use of *protection*? Condoms aren't perfect, but, if used consistently, they can reduce transmission of HIV by about 87 percent, and some studies have found them to be even more effective.

Although highly improbable, oral sex can also be a transmission method for HIV. A new study suggests that the tonsils, which are especially rich in immune cells (that some think can facilitate HIV infection), may provide a pathway for the virus to infect someone who's giving their male partner oral sex. Of course, the authors of the study are quick to point out that removing people's tonsils doesn't make sense as an HIV protection measure, when there are much less drastic and more affordable options available—condoms.

The more sexual partners one has over the course of a lifetime, the more risk of infection climbs. Now, a new study conducted by Dr. Adaora A. Adimora and colleagues at the

University of North Carolina, Chapel Hill, suggests that having multiple partners *at the same time* increases risk even further. In part because people engaged in multiple contemporaneous sexual relationships are more likely to engage in other risky behavior, such as drug and alcohol use during sex, and not use barrier contraception like condoms. The report, which was published in the *American Journal of Public Health*, also found that Hispanics and African Americans were as much as three times more likely as whites to have multiple sexual partners at the same time. Hispanics and African Americans are also more likely than whites to have HIV. "This study sheds light on the epidemic of heterosexually transmitted HIV in the U.S.," Dr. Adimora stated, "especially among African Americans and Hispanics."

It stands to reason that multiple concomitant sexual partners would increase the risk of HIV transmission. Scientists have long known that individuals are especially contagious when newly infected, especially during the phase of acute disease when they are experiencing those first flulike symptoms. So, if an individual becomes infected while in more than one sexual relationship, he or she exposes multiple individuals to infection. "People—especially women—need to avoid partnerships with people who have other partners," says Dr. Adimora.

Let us shatter the prejudicial myth once and for all. HIV is not just the province of homosexuals and swinging heterosexuals. Viagra and other drugs, along with better healthcare and social changes, have contributed to a sort of senior sexual revolution—more and more elderly people are having more and more sex. And with sex comes sexually transmitted infections. Sure enough, along with other STIs, HIV is on the rise among seniors. Tom Liberti, the chief of the Florida Bureau of HIV/AIDS, surprised the medical community when he reported that 16 percent of new HIV cases reported in 2005 were in people over fifty. I witnessed this senior sexual revolution first hand while doing my gynecology rotation. We had many patients, who being newly divorced, acquired HPV and then cervical cancer because they skipped using condoms. Unfortunately for these women they were thinking, being menopausal, that they couldn't get pregnant, and completely forgot about STIs.

The rising rate of HIV in seniors poses a whole new challenge for health-care professionals and other advocates promoting safer sex. Many elderly people today grew up in the pre-HIV era, before safer sex campaigns and so don't see the point in protection. And even if they do, they figure they'll die of old age before they die of AIDS. Of course, it's not right to assume that HIV won't cause additional health problems. Jolene Mullins works for the Broward County Florida Health Department on their Senior HIV Intervention Project. As Mullins points out, "The virus attacks the immune system and your immune system naturally breaks down with aging. If HIV is put on top of that, it naturally enhances the problems."

It all comes down to the same thing: infections don't discriminate. So whether you're gay or straight, old or young, having sex in the missionary position, orally, anally, or any other way—unless you are in a long-term monogamous relationship and you know your partner has a clean bill of STI health—be safe.

jagged little pill

One of the side effects of nature's decision to encourage reproduction by making the reproductive act so pleasurable is that lots of people are interested in sex for its own sake, and not for evolution's. The technologically sophisticated society in which we live has led to the development of many methods to foil nature's reproductive aims. Of course, it didn't take the advent of modern technology for humans to find ways to prevent pregnancy: contraception actually has a long and varied history around the globe; modern science and technology have just made it more effective. And it actually didn't even take humans to find ways to prevent pregnancy—new research has shown that contraception, maybe deliberate, but often accidental, occurs in the animal kingdom too.

■ ■ ■

CHEMICAL CONTRACEPTION—LIKE THE famous
birth-control pill—involves the ingestion of female hormones,
or chemicals that mimic them (more in a moment), to essentially trick the body into thinking it's pregnant. When a
woman is pregnant, she doesn't usually ovulate. If she doesn't
ovulate, she can't get pregnant.

Animals, of course, are not lining up at the local animal
pharmacy to fill their prescriptions. Instead, they unintention-
ally eat specific foods that contain either phytoprogestogens or
phytoestrogens—which are naturally occurring chemicals that
mimic female sex hormones. Why would plants be producing
chemicals that can mimic human or mammalian hormones?
The simple answer is, it's their way of quietly defending themselves, keeping their predators under control by reducing their
fertility. When animals eat enough phytoprogestogen- or
phytoestrogen-rich food, it's as if they're taking a natural
oral contraceptive pill. So which animals line up at the birth
control buffet? And, more to the point, why?

James Higham, a research fellow at Britain's Roehamp-
ton University, studied two troops of olive baboons living in
the rain forest of Nigeria's Gashaka-Gumti National Park.
They discovered that levels of phytoprogestogens in the waste
product of the troop's females climbed sharply every year from
August to October. They also identified only one food that
was eaten by both troops during that time of year—*Vitex doniana*, the African black plum and its leaves. And, sure enough,
laboratory tests of the black plum's chemical makeup revealed
it is chock full of phytoprogestogens. The scientists' study of

the troops revealed that all that phytoprogestogen consumption had a visible effect on female fertility. Unlike humans, whose bodies don't visibly change during ovulation although as we discussed some women may dress and behave differently during their most fertile period, many animals display obvious physical evidence that they're ovulating. In olive baboons, their hindquarters swell and redden, and that's what draws males to mate with them. But, during the months when the troops feasted on African black plums, that sexual swelling was markedly reduced. The combined internal and external effects of all those plums led the researchers to conclude that the plums pulled double duty as a contraceptive, preventing the physical cues that led to mating and hampering ovulation. The plum "appears to act on cycling females as both a physiological contraceptive (simulating pregnancy in a similar way to some forms of the human contraceptive pill) and a social contraceptive (preventing sexual swelling, thus reducing association and copulation with males)," write the authors.

Why the baboons go on "the plum" every year is still subject to speculation. Dr. Higham himself told *the Times of London* that he couldn't determine whether the baboons were deliberately dosing themselves with the contraceptive fruit or not. But one primate expert, Patricia Whitten of Emory University in Atlanta, Georgia, said she thinks the baboons may be inadvertently eating the plums to prevent disease transmission. August to October is the rainy season in that part of Nigeria, which also makes it the disease season. By eliminating copulation, the baboons are reducing the possibility of catching a contagious disease from another baboon through sexual intercourse.

If the baboons are intentionally going on "the plum" (which at this point seems highly unlikely), then this is the first evidence of an animal (besides humans, of course) deliberately consuming a food product that has contraceptive properties. There are other examples of animals eating foods that impede their reproduction; however, they just tend to be defense mechanisms of the dietary product in question. In the 1940s, agricultural specialists in Australia discovered that the cause of the sheep-breeding crisis they faced was the European clover their sheep were grazing on. Red clover produces a compound called formononetin that is converted into a powerful estrogen mimic by microorganisms in the guts of the animals that eat it. When a crop of clover is struggling due to bad weather, insects, or, in this case, transplantation from Europe to the drier Australian climate, it kicks up its production of formononetin. What does that do? It cuts down on the next generation of animals that would otherwise overwhelm it with grazing by sterilizing their potential parents. It's almost as if the plants were producing their own Pill as a defense mechanism.

Phytoestrogen consumption may not be the only way animal pregnancies are impeded. One controversial theory, called the *Bruce effect* (after British biologist Hilda Bruce, who first postulated it), suggests that exposure to unfamiliar males can terminate new pregnancies in female rodents of a variety of species. The Bruce effect has so far been observed only in the laboratory, which means some condition created by the laboratory environment itself may be entirely or partially responsible. Still, scientists have observed the Bruce effect in about a dozen different species, including domestic mice, deer mice, and voles. One study documented miscarriages in 88 percent of the female mice exposed to unknown males.

If the Bruce effect actually occurs in nature, scientists think it may be related to animal infanticide, which occurs with a fair degree of frequency among some animals, including lions, primates, and rodents. Typically, infanticide involves one or more dominant males who join a community in which fertile females have young children. The theory is that they kill the offspring to make their mothers sexually available to bear and nurture their own offspring. The Bruce effect may have evolved as an adaptation to male infanticide—instead of investing her energy and physical resources in bearing offspring that may later be killed, the female who encounters the unfamiliar male essentially cuts her losses. By aborting her current pregnancy, she can mate with the new male or males and will possibly have increased her future offspring's chance of survival.

Natural forms of contraception don't just occur in animals—humans experience them too. Immediately after a woman gives birth, and for some time afterward, she can be infertile; her body has yet to return to its normal cycle of ovulation and menstruation. Most people understand that you can't get pregnant the day after you give birth; what they may not realize is that regular breast-feeding can prolong this period of infertility by six months or more, through a phenomenon known as lactational amenorrhea. The technical term for failure to menstruate in a woman of normal reproductive age is amenorrhea. When this is caused by postpartum breast-feeding, it is called lactational amenorrhea. A study by the World Health Organization (WHO) found that nursing kept pregnancy rates down to just 1 percent during the first six months after delivery.

Lactational amenorrhea (LAM) doesn't occur with casual breast-feeding. It requires relatively continuous breast-feeding to create the condition with no supplemental feedings, such as formula or table food, for the infant. How it works is simple— the regular suckling of its mother's breast by an infant suppresses production of hormones necessary to trigger ovulation. No ovulation, no fertility.

Lactational amenorrhea is so effective that it is a recognized form of natural family planning that is even accepted by many religious groups opposed to contraception. Before you try this method at home, keep in mind that it can and does fail, especially the longer it's been since delivery. A year after birth, that same WHO study found that pregnancy rates climbed to around 7.5 percent. And there are some strict rules you have to follow in order to have reasonable confidence that it will work.

According to Planned Parenthood, using breast-feeding as birth control can be effective for six months after delivery only if a woman:

- Does not substitute other foods for a breast-milk meal
- Feeds her baby at least every four hours during the day and every six hours at night
- Has not had a period since she delivered her baby

If you're risk-averse and not opposed to contraception, the best way to keep from getting pregnant without hampering your milk supply (estrogen does that) is to combine LAM with the mini-pill, which is progestin only.

The evolutionary benefit of lactational amenorrhea is pretty straightforward. Providing a newborn with all the energy and nutrition it needs places a considerable strain on the body of a woman who is still recovering from the even more significant physiological stress of pregnancy. Adding a new pregnancy to the mix risks not only her ability to carry the new pregnancy to term, but also her ability to secure the success of the new child she has already invested so much in. Not to mention her own health. Since most newborns have possibly not consumed anything but breast milk for most of human history, humans have been using the lactational amenorrhea method for millions of years, however inadvertently.

THERE'S ANOTHER WAY that nature blocks fertility in human women. It happens at the other end of their development—menopause, which takes its name from the Greek words for *month* and *halt*. Menopause is a retrospective diagnosis, a year without menstruation in an older women, at which time her reproductive system ends its active period. Typically, menopause occurs in midlife—the average age in western countries is fifty-one. Why does female fertility shut down in midlife while male fertility continues unabated? Well, for most animals, it doesn't.

A recent study led by Harvard University postdoctoral student Melissa Emery-Thompson followed six populations of chimpanzees and compiled data about their fertility patterns and compared them to similar data from human hunter-gatherer groups. They found that humans and chimps both entered a period of reproductive decline in their forties—but that period of time coincides with normal chimp life expectancy, while humans live on for decades. In fact, female chimps that do live longer had no trouble reproducing well past forty. "Females in the wild and in captivity have given birth in their 50s, and the oldest living captive female, who is about 69, gave birth past the age of 60," said Emery-Thompson.

In other words, our primate cousins don't seem to experience menopause. "Human life history is in fact one of the most radical departures from the apes," Emery-Thompson explained. "We live longer than expected for our size, we have vastly higher reproductive costs, yet manage to reproduce much faster, we mature very slowly, and we have this peculiar post-reproductive period that distinguishes us from most other mammals."

So why shut down human fertility midstream? There are lots of theories out there. One is simply that the older a woman is, the harder it is for her body to weather the demands of pregnancy and childbirth. And, since newborns need help to survive, there could be evolutionary pressures against late-life pregnancies that put babies at risk of being orphaned at birth or shortly thereafter.

Another theory is the genetic shelf life of eggs. Men make fresh batches of sperm all the time, but it's thought that for the most part women are born with most of their eggs already in place. The genetic material in those eggs ages along with the rest of a woman's body, so some researchers believe menopause may prevent the risk of transmitting genetic errors resulting from aged and damaged eggs. Evidence to support this view can be found in the fact that genetic errors in embryos and fetuses climbs rapidly once a woman passes the age of thirty-five. Take Down syndrome, for example, a congenital condition usually caused by an extra copy or part of chromosome 21. At twenty, a woman's risk of having a child with Down syndrome is one in two thousand; at forty-nine, it is as high as one in twelve. There is some research that also implicates a father's age in Down syndrome, although it's still controversial and doesn't seem to have anywhere near the same strength as the maternal effect.

Still other researchers think there's no real mystery. For most of history, they say, life expectancy and the age of menopause were about the same.

And then there's the grandmother theory, a very interesting idea that's gaining traction. Essentially, according to this notion, by creating nonchildbearing grandmothers, menopause creates a team of unburdened additional caregivers: without children of their own, they can help their children, especially their daughters, raise the next generation.

A study led by Daryl P. Shanley of Newcastle University in the United Kingdom published in 2007, gives more weight to this theory. Shanley and his colleagues studied birth and death records of more than five thousand people in Gambia between 1950 and 1975 to see if they could find any grandmother effect. And they discovered some pretty powerful evidence: children who lost their mothers before they turned two were twice as likely to survive if their maternal grandmother were still alive. "Our results point clearly towards the maternal grandmother having a key role in the evolution of the menopause," said Shanley. Adding a little more weight to this theory from the perspective of overall cost and benefit to society, Emery-Thompson notes that grandmothers in hunter-gatherer societies bring in more calories than they eat.

Why maternal grandmothers in particular? Well, there's the obvious connection: maternal grandmothers were obviously pregnant with their daughters, and their daughters were obviously pregnant with their grandchildren. So they *know* they're related to their daughter's children. Given that men and women can both be unfaithful, and have been so throughout history, before genetic testing the only way to be absolutely certain you were related to somebody was through the maternal line. There's another way to test paternity-maternity; it's anything but accurate but in some cases it can work. If both parents don't have a widow's peak, a hairline that comes to a point midline and a trait that usually trumps any other hairline, then through a little genetic trick called dominance all their resulting children should not have widow's peaks. If the children do have widow's peaks, then most likely their mother cheated. In some cases this is a very visible way to discern fidelity.

But research also shows that the special connection between grandchildren and their maternal grandmothers is a two-way street: the children tend to be closer to their mom's mom too. I call this curious phenomenon "Darwin's grandma," and I think it goes beyond the obvious. Way beyond, in fact—all the way down to a tiny bit of cellular machinery called *mitochondria*.

Mitochondria have many functions, but their most prominent job is to provide energy for the cells they are part of, which is why they are sometimes called the workhorses of the cell. Mitochondria are fascinating in many ways, but we're concerned with one way in particular. Mitochondria have their own DNA—and unlike all your other DNA, in almost every cell in your body, mitochondria (except for rare exceptions) come from your mother alone. Which means you share mitochondrial DNA with your mother and with her mother and even her mother's mother, on up the line, but almost never with your father or anyone on your father's side. So, in a very real way, you have a greater biological connection to your maternal grandmother than your paternal one. This may explain, if you've ever wondered, where all that extra maternal grandmother attention may have been coming from.

THE HISTORY OF human contraception goes well beyond the natural protection of hormone suppression triggered by breast-feeding or the evolutionary development of menopause. In *Contraception and Abortion from the Ancient World to the Renaissance*, John Riddle describes thousands of years of human efforts to engage in sex while avoiding pregnancy. Some people used natural methods still practiced today, like *coitus interruptus*, the dubious practice of halting intercourse before ejaculation.

The rhythm method is the timing of intercourse to avoid the days in and around ovulation when women are most fertile. The rhythm method relies on the detection of a small drop in vaginal or rectal temperature that occurs twenty-four to thirty-six hours before ovulation, which is then followed by an abrupt rise of 0.5 to 0.7°F.

Three days after the temperature spike, a woman is no

longer in her most fertile period. By tracking temperature daily, it's possible to time intercourse around ovulation. Of course,
the margin for error is pretty slim.

Other women even track the consistency of their cervical
mucus in a natural method called the *Billings method*. You may
remember cervical secretions are usually thick and opaque and
somewhat spermicidal, but a few days before ovulation, they
become thinner, more watery, and somewhat stretchy. Think
mozzarella on a hot pizza, only transparent. This change is
thought to allow sperm better access through the cervix, and
to the egg, increasing the likelihood of pregnancy. Again, this
method can have some success, but it requires a certain degree
of skill and dedication.

The first barrier-type contraceptives seem to have involved questionable methods that are hard to imagine anyone
ever using—according to some accounts, ancient Egyptians
used crocodile feces, and ancient Arabs used elephant dung as
vaginal suppositories to prevent pregnancy. Riddle offers his
admittedly unscientific (but very understandable) reaction:

> A suggestion is made that feces may have actual birth con-
> trol properties, as an agent that either blocks mechanically
> the seminal fluid at the os of the cervix or changes pH level.
> In the absence of more and better evidence, this hypothesis
> represents too great a modern effort to impose scientific ra-
> tionality. A simple explanation—probably incorrect—is that
> inserting feces into a woman's vagina would be an excellent
> contraceptive merely by decreasing the libido of a squeamish
> male.

Primitive versions of artificial methods used today, such as oral contraceptives and condoms, followed. Oral contraceptives were first. According to Riddle, the earliest record of oral contraceptive use is found in an ancient Egyptian medical text now known as the Berlin Medical Papyrus, although we don't know exactly what was prescribed. The Berlin Medical Papyrus is one of a group of ancient Egyptian papyri that open a window onto the medical thinking and practices of the time. The Berlin Medical Papyrus was found in the vast Egyptian burial ground known as Saqqara, and is believed to have been written around 1300 B.C.E.

Some fourteen hundred years later, the Greek physician, herbalist, and pharmacologist Pedanius Dioscorides wrote *De Materia Medica Libri Quinque* ("Concerning Medical Matter in Five Volumes"), which was the *Physicians' Desk Reference* of its time—and of the next sixteen hundred years, actually. *De Materia Medica* is one of the few works by Greek and Roman scholars that never fell out of use during the Middle Ages. In it, Dioscorides suggests the use of white willow and juniper root to prevent pregnancy. And there is some modern evidence that both "prescriptions" may have achieved some success. White willow has been shown to contain estriol, a type of estrogen. Estriol is actually produced in extremely high quantities in women who are pregnant. White willow, which contains estriol, might have helped to prevent pregnancies by stopping women's bodies from ovulating. And juniper root has been shown to prevent embryos from implanting in the uterus of rats. Of course, the possibility exists that some of these herbal remedies worked at least some of the time. As Riddle put it, "Whatever can be written about the hailed placebo effect does not apply to birth control measures."

Andrea Tone, the author of *Devices and Desires: A History of Contraception in America*, cites Gabriel Fallopius as the first to describe a condom. If his name sounds familiar, it's because Fallopius, a sixteenth-century Italian physician and anatomist, gets credit for the discovery of the Fallopian tubes. In a paper on syphilis published in 1564, Fallopius urged the use of "a linen sheath" that had been soaked in an herbal bath to prevent transmission of the disease. The leap to pregnancy prevention was not far behind. As Tone writes:

> It was not long before people recognized that what prevented sexually transmitted disease probably prevented pregnancy, too. In the early eighteenth century, condoms made from oiled silk, fish bladders, and the intestines of goats, lambs, sheep, and calves were bought and used as contraceptives, making condoms the first modern birth control to acquire commercial validity.

As STIs spread across Europe, condom use became more and more widespread. Even Giacomo Casanova used condoms, supposedly calling them "English riding coats."

> Casanova wore the finest condoms money could buy, but he was hardly enthusiastic about them. It was emasculating and dispiriting, he complained to have to "shut myself up in a piece of dead skin in order to prove I am perfectly alive." Casanova's complaint, a familiar refrain among condom-wearing men then and now, did not prevent him

from promoting condoms or wearing them during frequent visits to French brothels. The condom, he announced, was a "wonderful preventive" for "shelter[ing] the fair sex from all fear."

Not everyone credits Casanova with such pure motives. British biographer and actor Ian Kelly thinks that Casanova wore condoms made from "rendered sheep gut" and "he used them mainly with nuns, who were particularly worried about getting pregnant."

The first rubber condom was made in 1855, and the first latex condoms—thinner, more pliant, and stronger than their rubber cousins—were produced in the 1930s.

Latex condoms weren't the only contraceptive developments of the nineteenth and twentieth centuries, of course—not by a long shot. A German gynecologist named Friedrich Wilde made custom-molded rubber cervical caps for some of his patients. By covering the cervix and blocking the entrance to the uterus, these caps prevent sperm from reaching an egg. The diaphragm also works by blocking the cervix, but, instead of sitting atop it like a cap, it is a soft dome that has a spring around the rim that creates a seal with the vaginal wall. The first diaphragm is credited to another German gynecologist, C. Haase, who described it in papers published under the pseudonym Wilhelm P. J. Mensinga in the 1880s. One of the benefits of using a diaphragm is that it can be inserted up to six hours ahead of time, which leaves room for spontaneity. And it is reusable. Like a cervical cap, diaphragms need to be fitted by a health-care professional to account for normal anatomical size differences between women. It's generally recommended that diaphragms be used in conjunction with a spermicide to increase their effectiveness.

The next leap in condom technology happened in the 1990s, when the first polyurethane condom was introduced. Many users found polyurethane condoms better than traditional latex because they conduct heat better, which makes wearing them more pleasurable. On the other hand, polyurethane is not as pliable and is more prone to breaking, which can defeat the purpose. Then there's the female condom, also made of polyurethane, but with two flexible rings at either end. It's inserted into the vagina before sex. But many people don't like them because they're somewhat bulky. Think baggy as opposed to fitted jeans.

The most popular insertional device of all is the intrauterine device, or IUD. In fact, IUDs are the most popular method of birth control in the world besides sterilization. They are used by more than 150 million women, a great percentage of whom are in China. Unlike other reversible contraceptives, IUDs must be inserted—and removed—by a health-care professional, but require no other action by the user until they need to be replaced, which is generally between five and ten years from insertion.

The first IUD sold to the public was made by G-spot namesake Dr. Ernst Gräfenberg in 1929. Early IUDs had significant rates of infection and expulsion. Today, there are two types of IUDs available in the United States. The first type contains copper (its technical name is T380A), and the second type releases hormones. Both use one or more small threads leading from the bottom of the IUD, where they protrude slightly from the cervix into the vagina, allowing a woman to check periodically and make sure it's in place. No one really knows how the copper IUD works. It's thought that it may prevent fertilization by its simple presence—the body reacts to it more or less as an invader, which creates an unfavorable environment for conception and implantation. As long as the IUD is present in the uterus, the possibility of fertilization is exceptionally small, less than 1 percent. One of the major benefits of using this type of IUD is lifespan and low maintenance. Once inserted properly, it can be left in place for up to ten years. Overall, the IUD is widely regarded as safe today, but that hasn't always been the case. For a long time, IUDs were thought to increase the risk of ectopic pregnancies. Because IUDs make uterine pregnancies so unlikely, it was thought that IUDs were actually *causing* the ectopic pregnancies, which doesn't appear to be the case. The IUD stops pregnancies from progressing *once a fertilized egg reaches the uterus*, by preventing implantation. But, if a fertilized egg implants before it gets to the uterus, as in the case of ectopic pregnancies, the IUD has no effect on it.

There were, however, real problems with some older IUDs. One of the most infamous devices in medical history, the Dalkon Shield, was a plastic IUD made and sold by the A. H. Robins Company in 1971. At the time, the U.S. Food and Drug Administration did not require testing and approval of medical devices in the same way required for pharmaceuticals. A. H. Robins conducted only one test of the product before taking it to market, a year-long trial to examine its efficacy as a contraceptive device that was led by the product's own inventor. And senior executives of the company were aware of a flaw in the Dalkon Shield's design that gave it a propensity to "wick" bacteria from the vagina into the uterus.

Which is exactly what it did—to thousands and thousands of women. "In addition to being responsible for at least eighteen deaths, the Dalkon Shield caused over 200,000 infections, miscarriages, hysterectomies, and other gynecological problems and led to an untold number of birth defects, caused by contact between the device and the developing fetus," writes Tone.

A few hundred thousand people filed lawsuits because of problems caused by the Dalkon Shield, resulting in one of the biggest class action lawsuits of all time, with a $2.5 billion settlement. In the category of silver linings, Congress passed a law in 1976 giving the FDA the authority to require testing and approval of medical devices, and President Ford quickly signed it, saying it "eliminate[d] the deficiencies that accorded the FDA 'horse and buggy' authority to deal with 'laser age' problems." Thankfully, today's IUDs don't seem to suffer from the same problems as their ancestors and are a good contraceptive choice, especially in women who have already had children.

Another doctor-assisted, maintenance-free contraception is sterilization—but it isn't really considered reversible, although it can sometimes be reversed through additional surgery. In women, the procedure is called tubal ligation and involves severing, scarring or clipping both Fallopian tubes. In males, the equivalent procedure is called a vasectomy and involves severing the vas deferens, preventing the transit of sperm from the testicles.

I was first exposed to vasectomies while I was working for the Population and Community Development Association, a Thai NGO. The founder of PDA, Mechai Viravaidya, is widely known as Thailand's Condom King for his work in making condoms more culturally acceptable. I lived across the street from PDA's Cabbages and Condoms restaurant, which, besides having an admirable collection of condoms from all over the world, helps provide some financial support for the organization. As part of its mission, PDA also advertised "non-scalpel" vasectomies. The technique is a lot less innovative than it might sound. They used scissors instead, which worked, of course, but it certainly wasn't the medical miracle "non-scalpel" seemed to promise. Their big innovation was to use a converted Winnebago to perform vasectomies around the country, which they offered for free on Father's Day.

When it comes to fame and impact, no other form of contraception can compare to the Pill, or the combined oral contraceptive pill, if you're being formal. In the combined form the Pill actually delivers derivatives of two steroid hormones, estrogen and a synthetic form of progesterone usually referred to as a progestin. For women who don't want to remember to have to take a pill every day, there's a patch that only needs to be changed every three weeks and delivers its hormones through the skin. There are also versions of progestin-only pills, usually called mini-pills, and medroxyprogesterone acetate, sold under the trade name Depo-Provera, an injectable progesterone that works for up to three months a shot.

The birth control pill has many parents, but the chemical key to its own birth was the development of the first synthetic progestin, norethindrone, which was synthesized by the master chemist, Carl Djerassi, along with a young Mexican chemist named Luis Miramontes in 1951. (I had the honor of meeting met Djerassi just as I was starting college. For an aspiring physician-scientist fascinated by medicine and chemistry, it was a bit like meeting the lead singer of your favorite band.) Djerassi and Miramontes had received an assist a decade earlier when a chemist named Russell Marker found a natural progestin called disogenin in the Mexican wild yam. Before Marker found disogenin, development of the Pill was stymied because all the chemicals being used to develop it were so expensive. Mexican yams, not so much.

Before we go further in describing the Pill's creation, it's worth taking a moment to explore a little-known but somewhat darker side to its development. A year before Djerassi's breakthrough, birth control advocate Margaret Sanger and philanthropist and women's rights advocate Katharine McCormick teamed up to encourage the development of a medical contraceptive that would give women individual control over their own fertility. Sanger was the founder of the American Birth Control League, the precursor organization to Planned Parenthood. Although it's not widely promoted, her strong advocacy of birth control was not only rooted in a conviction that women deserved control over their individual reproductive decisions. Sanger also believed society needed to exert control over who exactly reproduced in order to prevent reproduction in families that were genetically "unfit."

This philosophy—that society ought to exert an active control in determining who has children and who does not, in order to improve the gene pool—is called eugenics, and it has a terribly checkered history. Eugenics is generally divided

into two types: positive eugenics is the promotion of reproduction among those considered genetically well-off; negative
eugenics is the opposite, the intentional discouragement of reproduction among those deemed somehow less genetically fit.
Sanger was an outspoken proponent of negative eugenics. In
one pamphlet, she wrote:

> It is a vicious cycle; ignorance breeds poverty and poverty
> breeds ignorance. There is only one cure for both, and that is
> to stop breeding these things. Stop bringing to birth children
> whose inheritance cannot be one of health or intelligence.
> Stop bringing into the world children whose parents cannot
> provide for them. Herein lies the key of civilization.

To be fair, Sanger was by no means among the most radical eugenicists. She explicitly rejected the extreme—and hor-
rible—ideas of people like William Robinson, who proposed
killing the children of those deemed "unfit." She publicly
criticized the anti-Semitism of Nazi Germany, which based
its plan for creation of a "master race" on eugenics. And, ultimately, she believed that individuals, not the government,
must have the power to make their own reproductive decisions, which squared nicely with her support for development
of female contraception.

> The campaign for birth control is not merely of eugenic value,
> but is practically identical with the final aims of eugenics. . . .
> We are convinced that racial regeneration, like individual re-
> generation, must come 'from within.' That is, it must be au-
> tonomous, self-directive, and not imposed from without.

Katharine McCormick's motives don't seem tainted by the eugenic influence. A lifelong advocate of women's rights, beginning with the suffrage movement, she also had a keen interest in science. She was the second woman to graduate from the Massachusetts Institute of Technology and the first to do so with a degree in science. In 1906, her husband, Stanley McCormick, an heir to the American harvester fortune, was diagnosed with severe schizophrenia—a disease that also plagued his sister. Tone writes: "his sufferings, combined with her fears that schizophrenia could be inherited, forged in Katharine a resolve to stay childless and made her an early convert to contraception." McCormick and Sanger met in 1917 and began a long collaboration to give women individual control over their own fertility.

In 1951, the same year that Djerassi had his synthesizing breakthrough, Sanger met a gifted biologist named Gregory Pincus, who had been studying hormones and fertility for a quarter century. Sanger encouraged Pincus to begin research into an oral contraceptive for women and secured a small grant to help him begin. Sanger introduced McCormick to Pincus and his work in 1953, and she provided him with a massive increase in funding that set the stage for the ultimate development of the first oral contraceptive. The Pill was approved by the FDA in 1960 and was the most popular form of birth control in the United States by 1965. As Tone describes, it took the country by pharmaceutical storm:

> It was one of the greatest inventions of the twentieth century, the capstone of decades of pharmaceutical research. It inspired songs, cartoons, political debate, and grateful

letters from women around the world who flocked to their physicians' offices for prescriptions. The Catholic Church condemned it as immoral, and several African-American leaders denounced it as technology of genocide. . . . But no matter how Americans felt about it, the object of excitement needed no special introduction. By the mid-1960s, Americans knew the wonder drug of the decade simply as "the Pill."

When the Pill is prescribed as it was originally marketed, and usually still is, women take the Pill every day for twenty-one days and then either take a week off or take a placebo pill (one with no active ingredient) for seven days. During the week in which a woman is not taking the Pill, but taking the placebo or nothing, she will experience a period of bleeding similar to and occasionally lighter than normal menstruation. Many people assume this period of bleeding *is* menstruation, believing women go off the Pill so they can have their period. In fact, the bleeding is not menstruation per se; it's actually a symptom of withdrawal from the hormones. The Pill is sold this way because its original marketers assumed women would be more receptive to a product that preserved the appearance of a normal menstrual cycle along with its monthly reminder and confirmation of a lack of pregnancy.

If a woman were to take daily doses of the actual Pill, she would live a period-free existence. And many doctors have quietly advised patients who experience especially painful periods to do exactly that. In 2003, Barr Pharmaceuticals introduced Seasonale, a version of the Pill prescribed in such a way to reduce a woman's period from once a month to just four times a year. The trick to Seasonale is essentially all in the packaging—instead of twenty-one active pills and seven placebos, Seasonale is sold in packs of ninety-one pills, containing eighty-four active pills and seven placebos.

Of course, the obvious question is, why stop there? If you can sell a version of the Pill that eliminates eight periods a year, why not go for broke and sell one that gets rid of them entirely? That's what drugmaker Wyeth decided to do with its new version of the Pill, called Lybrel, that was approved by the FDA in 2007. It's not an instant end to bleeding—Wyeth's own website advises that breakthrough bleeding is common as women adjust to the constant hormonal treatment, and warns, "When prescribing Lybrel, the convenience of having no scheduled menstrual bleeding should be weighed against the inconvenience of unscheduled breakthrough bleeding and spotting." The biggest problem with Lybrel is that, because women don't menstruate regularly, if they do become pregnant they may not know it.

Breakthrough bleeding and spotting are actually the most common side effects of oral contraceptives, although as side effects go, they fall in the inconvenient but relatively inconsequential category. In many women, the Pill helps to reduce acne, which is generally considered a benefit, and is sometimes linked to larger breast size: Tone notes that the sale of C-cup bras in the United States climbed by 50 percent from 1960 to 1969.

The Pill has long been rumored to cause weight gain, although the evidence is less than conclusive. British researcher Dr. Sunanda Gupta conducted an extensive review of medical literature regarding weight gain and the Pill. Her conclusion? No evidence. And the report also suggested that rumors of this side effect had a side effect of their own—adolescent pregnancy. Gupta believes adolescent girls' fear of weight gain discourages many of them from using oral contraceptives, contributing to Britain's high rate of teen pregnancy.

In terms of serious side effects, the combination Pill is associated with some risk of cardiovascular disease, especially blood clots that can lead to heart attack or stroke. Those risks are magnified in women who smoke. The mini-pill (progestin only) is thought to be somewhat safer for women smokers. There is also some evidence that the Pill may be associated with an increased risk of depression. But taking the Pill is not all bad. Multiple studies have produced evidence that oral contraception use can lower rates of endometrial cancer (some studies found protection from ovarian cancer as well, but this is still controversial). Progestin, the synthetic version of progesterone that is one of the combined Pill's two main ingredients, has been shown to contribute to a decrease in serotonin levels in the brain. Serotonin is thought to help to regulate mood and emotional well-being; low levels of serotonin are linked to depression. Prozac and other similar drugs in its class, selective serotonin reuptake inhibitors, or SSRIs, help to treat depression by preventing serotonin's reabsorption into neurons. This boosts the overall serotonin levels in the synapses, between neurons in the brain. A recent study by Dr. Jayashri Kulkarni of Australia's Alfred Psychiatry Research Center found that women using the Pill showed significantly more symptoms of depression than a matched group who were not using the Pill. More research is clearly in order—but if you feel that a bad case of the blues descended on you when you started taking the Pill, talk to your doctor about it.

But the most unexpected side effect of the Pill may involve its playing chemical havoc in the one area it's supposed to help—your love life. Remember the T-shirt study run by Claus Wedekind that was the first to reveal that women were more attracted to the smell of men whose HLA genes (the genes that code for a critical part of our immune systems) differed from their own? The basic rule is this: when it comes to body odor and HLA, opposites don't just attract, they can apparently smell better too.

With one big exception.

Beginning with the initial Wedekind research, a few studies have shown that the Pill seems to shift women's olfactory preferences (in men, that is) into reverse. Instead, of preferring men with *dissimilar* HLA, they prefer men with *similar* HLA. Now, remember that HLA diversity—having dissimilar HLA between partners—a marker for genetic variability between the couple is associated with better fertility and gives your offspring a better chance at a stronger immune system. That's why scientists believe women are naturally attracted to the smells of men with somewhat dissimilar HLA; you're more likely to reproduce and more likely to produce children with strong immune systems, two big pluses from evolution's point of view.

A very recent study published in 2008 by Craig Roberts and colleagues at the University of Newcastle in the United Kingdom examined the smell preferences of a group of women before and after they began using the Pill. They also compared those results to a control group of women who were never on the Pill. There was one clear conclusion, as the authors wrote, "Across tests, we found a significant preference shift towards MHC [HLA] similarity associated with pill use, which was not evident in the control group."

If the Pill throws you off the scent, it could actually have a dramatic impact on whether Mr. Right is actually right for you. Dr. Dustin Penn, director of the Konrad Lorenz Institute for Ethology in Vienna, Austria, said, "It wouldn't surprise me if sabotaging our reproductive machinery would lead to faulty mate choice." According to Roberts and coauthors, "We do not know whether the change in preferences related to pill use is sufficiently strong to influence partner choice, but it could do so if odor plays a significant role in actual human mate choice."

And psychologist Rachel Herz, author of *Scent of Desire*, reiterated to me that if the Pill contributes to a woman choosing a mate with similar HLA, "it's like picking your cousins as marriage partners. It constitutes a biological error." So how does the Pill point your nose in the wrong direction?

As we discussed, by elevating the level of hormones circulating through a woman's body, the Pill essentially tricks her body into thinking it's pregnant, preventing ovulation. No ovulation, no chance of pregnancy. Of course, as everybody knows, pregnant women experience a huge array of changes, and we're not just talking about the bump or a sudden predilection for pickles. Wedekind thinks women's odor preferences may change during pregnancy too. Once a woman is pregnant, she instinctively looks to find the safest environment for her unborn child—family, of course, is deeply associated with safety. So it's possible that, when pregnant, a woman is more attracted by the smell of family than the smell of a potential mate—she's sniffing for safety, not sniffing for sex.

So what's to be done about it? Herz has a straightforward recommendation: When it comes to women "who are currently trying to find the man that they want to have a family with, in this quest, prior to embarking on it, they should go off the Pill. And they should also try to subtly tell their men that they become involved with not to put on heavy fragrance because the fragrance can mask their body odor as well."

On the other hand, if you were on the Pill when you found the man of your dreams, according to Herz, don't worry about it. If you're in love, you're in love; it's too late. And guys, don't worry that your girlfriend, fiancée, mistress, or wife is suddenly going to hate the way you smell when she goes off the Pill (as long as you have a good relationship, anyway). As Herz says:

> Once you've fallen in love with someone, once the emotional attachment has been made to an individual, they could smell like a garbage truck and you'd be attracted to that smell. If you're healthy, and I've been on the Pill, and you've been wearing cologne, I don't really know how you truly smell. Once you stop wearing cologne, and you're sweating around the house, and now I know how you really smell because we're past that stage of courtship, then how you smell now is going to be associated with how I feel about you. And if I'm in love with you, then that's going to be how I associate the meaning of that smell.

In other words, a rose by any other smell will smell as sweet.

On the other hand, it's possible that the havoc the Pill plays on smell preferences may be contributing to the high rates of divorce in developed countries where Pill use is prevalent. As long as you're in love with someone, they're going to smell good to you. But if you fell in love with someone while you were on the Pill, and you're having relationship difficulties now that you're off the Pill, it's possible the situation is adding olfactory insult to relationship injury. And given how repulsive bad odors can be, it's possible that this change in smell perception makes it very difficult for a couple in this situation to reconcile. Herz believes:

> If I now smell you differently given that I'm not on the Pill and I don't like you much anymore, then your smell is going to become highly offensive. If someone smells bad to you, the act of being intimate with them sexually is, I think, next to impossible. I have wondered whether or not the reason for the consistency of this complaint [about their partners' smell] and, potentially, the reason for this high, high rate of divorce, has to do with having been on the Pill, then going off the Pill, and the meaning of the man's scent changing and, therefore, this sort of revulsion setting in and no longer being able to be with the person.

None of this is to say that the Pill is going to destroy your relationship. But it would be nice if there were oral contraceptive options that didn't interfere with a woman's olfactory radar, right? Well, guess what may one day be coming to a pharmacy near you?

A male pill. Well, it might be a patch, a gel, or an injection, but you get the idea.

A series of clinical trials has now demonstrated that a mix of hormones administered to men can—reversibly—halt production of sperm. The studies show that normal sperm production resumes in about three months. Most of the formulations under research involve the administration of testosterone together with progestin in one form or another. Some people believe that implanting the drug might make it more effective, which would also give men a "badge of honor" allowing them to certify to their current partners that they're actually not currently fertile. There's still a lot of work to be done, but scientists are getting closer. The latest efforts target the testicles directly, which is where sperm are produced. And if that doesn't work, there's always soy.

Recent research published in 2008 by a team from Harvard seems to back up the idea that increased consumption of soy and related isoflavones can result in lower sperm counts in men. Soy contains many phytoestrogen compounds similar to those that kept those sheep eating European red clover from getting pregnant. Very importantly, men did not become sterile from soy consumption; their sperm counts just dipped down. Although it may lead to new avenues of future contraception research, it's still a bit premature for men to turn to tofu instead of condoms.

It's not clear when a reversible chemical male contraceptive will hit the market—but it's clear that it's coming. And that will give us one more opportunity to examine what happens when we try to pull the wool over nature's eyes. Or nose.

good vibrations

New developments in contraception are just one way scientific discoveries and technological breakthroughs are going to continue to change the way we live together, get together, and have sex. Advances in disciplines as diverse as chemistry, robotics, and communications are multiplying the possibilities for sexual interaction, just as they're changing the way we live our daily lives. There's one big difference between technology's effect on sex and its effect on everything else. You can always go back to the good old days. It's pretty hard to live in the twenty-first century and avoid electronic communication in some form or another, for example. But when it comes to sex, with its relatively easy mode of use, the old-fashioned way will never go out of style.

As we've discussed throughout the book, the intersection of

technology and sex isn't actually new, either: humans have been using technology, in the form of cosmetics and herbal medicine and all kinds of mechanical devices, to sex up our sex lives for thousands of years. There are written records of oral contraceptives that go back more than two thousand years and written records of condoms that go back more than five hundred years. Humans turned to surgery to enhance or fix their sexual appeal centuries before breast implants made their controversial debut. Even the vibrator—the most popular sex toy in the world today—is already about a century and a half old.

None of which should come as much of a surprise, given how important sex is to human life. More or less a prerequisite, in fact. So it makes sense that as long as we've had the capacity to innovate, we've been innovating our sex lives. In fact, it's this capacity to take control over our sex lives—and so many other aspects of our lives—that really sets us apart from other animals.

Evolution's twin imperatives—survival and reproduction—have driven the development of sex in every species that has it. Evolution is behind the development of those species that reproduce asexually too, of course. For starters, remember that sexual reproduction may have evolved as a double-barreled mechanism that allows each generation the best possible chance for survival. It also gives parents a chance to spare their offspring a nasty inheritance of free-loading parasites, and it allows for a genetic reshuffling of the deck that can increase the survival odds of a species with the creation of new traits.

Of course, evolution isn't the type to just set the table and see who shows up for dinner. Which explains the development of sexual behaviors and rewards. In humans, it is *pleasure* which promotes bonding between couples. Evolution, in other words, will set the table, pick the flowers, pour the wine, and write the menu—over and over again.

All of this is as true for humans as it is for every other species that uses sex for reproduction. Evolutionary pressures to find the right mate (or mates) and reproduce with them are behind much of our sexual desires and behavior. The cross-cultural preference for hourglass-shaped women is connected to nature's never-ending preference for fertile, healthy partners with well-suited traits. Who knew that a woman turned on by a man's smell may actually be enjoying the alluring scent of his human leukocyte antigen, that key part of our immune system? That men may be more appealing to some women if they are darker, because having more pigment in their skin helps them block out UV sun rays that can lower folate levels, which means they have better sperm than their lighter-skinned brethren. Or that, if there is a genetic link to homosexuality in males, it may be a gene or a combination of genes that *make* heterosexual women especially attracted to men, encouraging them to reproduce more.

We also now know that evolution's obsession with reproduction may have left its traces in a penchant for infidelity. On the other hand, its equally strong fixation on survival may be found in the way sex reinforces the abiding love that brings and keeps two people together and helps them raise a family. Some researchers even believe that sex is responsible for the most uniquely human trait of them all—our big brains. Psychologist Geoffrey Miller, in his book *The Mating Mind*, argues that the development of our relatively large human brain may have evolved to meet the demands of court-ship and, thus, sex.

If he's right, that would really bring things full circle. Because it's that big brain that gives us a chance no other species seems to have. Every other species under the sun is governed by its sexual instincts and urges, but we don't have to be. We can *choose* to give in to all those evolutionary pressures and physical instincts, or override them, as challenging as that may seem sometimes. Evolution cannot make anyone unfaithful, and it's not responsible if you cheat. Instead, evolution has given us the power to be whoever and however we aspire to be. If you have an extra X- or Y-chromosome, are transgendered, straight, gay, or bi, it's your big brain that gives you the power to stay faithful. Just like it gives you the power to say no to another helping of cake or go to the gym or pay extra attention to your son's children, even if they didn't inherit your mitochondria.

And new technology is only going to increase our ability to take control of our lives, sex lives included. Cosmetic surgery is one way. Contraception is another. New technology can even help couples in long-distance relationships to maintain a sexual connection when they're apart. Sexual remote interaction technology—termed *teledildonics*—allows someone to remotely manipulate a vibrator being used by his or her partner through the Internet. And, of course, it's not all sexual fun and games. Technology can help us overcome the challenges of sexual reproduction too, giving us new insight and understanding into diseases of sexual development (or intersexuality) and better options to help people manage them. And fertility treatments, which have already given thousands of people with reproductive problems or sexual dysfunction the ability to have children, are constantly improving. We may even see the successful cloning of humans in the next few years.

That's the great power and gift of humanity. We are not simply ruled by our physical condition, even when it comes to sex, that most intimate and physical of acts. We can ask why, discover an answer, and then ask, what now?

Of course, the more we know, the more we realize how much we have to learn. Most doctors are completely unfamiliar with female ejaculation, for example, sometimes leading to all kinds of unnecessary procedures to "cure" an incontinence problem that doesn't exist. In the last few years, science has made some startling sex-related discoveries that have enormous potential to save lives—such as the new vaccine against HPV which has the potential to save millions of women from cervical cancer. We've learned how herpes can influence an individual's sexual behavior through "oral exploration," in order to facilitate its own transmission. We are reminded about the incredible interconnectedness of all life, as a seemingly minor trend in personal grooming—the Brazilian wax—may be pushing a species, the pubic louse, (admittedly a pretty unattractive one) toward extinction.

At the same time that we continue to push the frontiers of sexual science, we need to make sure that people are equipped with the knowledge that they need to have safer, healthier and more rewarding sex lives. It ought to be astonishing, for example, that one in five American high school girls doesn't know how HIV is transmitted. And millions of adults still don't know enough about how to get the most out of their relationship with their partners.

You want to know what the magic ingredient to a good sex life is? Understanding.

Realize that it may not be possible to completely escape evolution's grasp. So what can we do? Be open and adapt. Understand what you like, and why you like it. Learning about the influences that millions of years of trial and error have played in our evolution as a species can bring us closer to breaking free from instincts and make informed choices.

The more we understand how sex works, the greater the opportunity we have to enjoy one of evolution's greatest gifts.

notes

Chapter 1: Girls Just Want to Have Fun

For a long time, menarche: See K. Zhang, S. Pollack, A. Ghods, C. Dicken, B. Isaac, G. Adel, G. Zeitlian, and N. Santoro, "Onset of Ovulation After Menarche in Girls: A Longitudinal Study," *J Clin Endocrinol Metab* 93, no. 4 (2008): 1186–1194.

Something strange is affecting: M. G. Elder, *Obstetrics and Gynaecology: Clinical and Basic Science Aspects* (Imperial College Press, 2001); see also W. C. Chumlea et al., "Age at Menarche and Racial Comparisons in U.S. Girls," *Pediatrics* 111 (2003): 110–113. G. Chodick, A. Rademaker, M. Huerta, R. Balicer, N. Davidovitch, I. Grotto, "Secular Trends in Age at Menarche, Smoking, and Oral Contraceptive Use Among Israeli Girls," *Prev Chronic Disease* 2, no. 2 (2005): A12; X. Du, H. Greenfield, D. R. Fraser,

K. Ge, W. Zheng, L. Huang, and Z. Liu, "Low Body Weight and Its Association with Bone Health and Pubertal Maturation in Chinese Girls," *European Journal of Clinical Nutrition* 57 (2003): 693–700.

According to psychosocial acceleration theory: B. J. Ellis, "Timing of Pubertal Maturation in Girls: An Integrated Life History Approach," *Psychology Bulletin* 130, no. 6 (2004): 920–958.

Then there's a theory: Matchock, R. L., and Susman, E. J., "Family composition and menarcheal age: anti-inbreeding strategies," *American Journal of Human Biology*, 18(4) (2006): 481–491.

Another theory that has been: J. Lee, M. P. H. Appugliese, N. Kaciroti, R. Corwyn, R. Bradley, J. Lumeng, "Weight Status in Young Girls and the Onset of Puberty," *Pediatrics* 119, no. 3 (2007): 624–630.

One study suggests: W. D. Lassek and S. J. Gaulin, "Brief Communication: Menarche Is Related to Fat Distribution," *American Journal of Physical Anthropology* 133, no. 4 (2007): 1147–1151.

"This fat is protected: Ibid.

Whatever the biological cause: P. Gluckman and M. Hanson, *Mismatch: Why Our World No Longer Fits Our Bodies* (Oxford, UK: Oxford University Press, 2006).

Human breasts are unique: For additional information see the following study, F. E. Mascia-Lees, J. H. Relethford, and T. Sorger, "Evolutionary Perspectives on Permanent Breast Enlargement in Human Females," *American Anthropologist* 88, no. 2 (1986): 423–428.

The average size of breasts: V. Lambert, "Why the British Woman's Cleavage Has Gone from 34B to 36C in a Decade," *Daily Mail*, January 23, 2008.

Zoologist and bestselling author: D. Morris, *The Naked Woman: A Study of the Female Body.* (New York: Thomas Dunne Books, 2005).

Men also have breasts and nipples: N. Swaminathan, "Strange but True: Males Can Lactate," *Scientific American*, September 10, 2007; J. Diamond, "Father's Milk—Male Mammals' Potential for Lactation," *Discover*, February 1995.

Scaramanga: B. Howard, H. Panchal, A. McCarthy, and A. Ashworth, "Identification of the Scaramanga Gene Implicates Neuregulin-3 in Mammary Gland Specification," *Genes and Development* 19, no. 17 (2005): 2078–2090.

A third nipple is not only: L. S. Gendler and K. A. Joseph, "Images in Clinical Medicine: Breast Cancer of an Accessory Nipple," *New England Journal of Medicine* 353, no. 17 (2005): 1835.

Breasts aren't the only: R. W. Taylor, E. Gold, P. Manning, and A. Goulding, "Gender Differences in Body Fat Content Are Present Well Before Puberty," *International Journal of Obstetric-Related Metabolic Disorders* 21, no. 11 (1997): 1082–1084. N. Gungor, S. A. Arslanian "Chapter 21. Nutritional disorders: integration of energy metabolism and its disorders in chidhood" in Sperling MA, ed. *Pediatric Endocrinology* (2nd ed) Philadelphia, Saunders pp. 689–724.

But here's where: M. Rozmus-Wrzesinska and B. Pawlowski, "Men's Ratings of Female Attractiveness Are Influenced More by Changes in Female Waist Size Compared with Changes in Hip Size," *Biol Psychol* 68, no. 3 (2005): 299–308.

In the early seventeenth century: John Harrington, as quoted in R. Khamsi, "The Hourglass Figure Is Truly Timeless," *New Scientist*, January 10, 2007.

Other social research: B. Low, *Why Sex Matters: A Darwinian Look at Human Behavior* (Princeton, NJ: Princeton University Press, 2000).

In 1996, Harvard researchers: S. F. Lipson and P. T. Ellison, "Comparison of Salivary Steroid Profiles in Naturally Occurring Conception and Non-Conception Cycles," *Human Reproduction* 11, no. 10 (1996): 2090–2096.

A 2004 Polish study: G. Jasienska, A. Ziomkiewicz, P. T. Ellison, S. F. Lipson, and I. Thune, "Large Breasts and Narrow Waists Indicate High Reproductive Potential in Women," *Proc Biol Sci* 271, no. 1545 (2004): 1213–1217.

Devendra Singh, the psychologist: D. Singh, P. Renn, and A. Singh, "Did the Perils of Abdominal Obesity Affect Depiction of Feminine Beauty in the Sixteenth to Eighteenth Century British Literature? Exploring the Health and Beauty Link," *Proc Biol Sci* 274, no. 1611 (2007): 891–894. S. Bhattacharya, "Barbie-Shaped Women More Fertile," *New Scientist,* May 5, 2004.

Even more fascinating: W. D. Lassek and S. J. C. Gaulin, "Waist-Hip Ratio and Cognitive Ability: Is Gluteofemoral Fat a Privileged Store of Neurodevelopmental Resources?" *Evolution and Human Behavior* 29 (2008): 26–34

In the hierarchy of attraction: Many papers have been published on this topic. See, for example, A. C. Little, C. L. Apicella, and F. W. Marlowe, "Preferences for Symmetry in Human Faces in Two Cultures: Data from the UK and the Hadza, an Isolated Group of Hunter-Gatherers," *Proceedings of the Royal Society*, B, no. 274 (2007): 3113–3117; I. N. Springer et al., "Facial Attractiveness: Visual Impact of Symmetry Increases Significantly Towards the Midline," *Ann Plast Surg* 59, no. 2 (2007): 156–162; D. Perrett et al., "Symmetry and Human Facial Attractiveness," *Evolution and Human Behavior* 20, no. 5 (1999): 295–307; M. J. Tovee, K. Tasker, and P. J. Benson, "Is Symmetry a Visual Cue to Attractiveness in the Human Female Body?" *Evolution and Human Behavior* 21, no. 3 (2000): 191–200; S. Jasienska, P. Lipson, I. Thune, and A. Ziomkiewicz, "Symmetrical Women Have Higher Potential Fertility," *Evolution and Human Behavior* 27, no. 5 (2006): 390–400.

One of the more noticeable: G. Gallup, "Permanent Breast Enlargement in Human Females: A Sociobiological Analysis," *Journal of Human Evolution* 11 (1982): 597–601.

A British study published: D. Scutt, G. A. Lancaster, and J. T. Manning, "Breast Asymmetry and Predisposition to Breast Cancer," *Breast Cancer Research* 8, no. 2 (2006): R14.

Before we proceed: A few good resource books include: Boston Women's Health Book Collective, *Our Bodies, Ourselves: A New Edition For a New Era*, 35th anniversary ed. (New York: Simon and Schuster, 2005); C. Livoti and E. Topp, *Vaginas: An Owner's Manual* (New York: Thunder's Mouth Press, 2004); N. Angier, *Woman: An Intimate Geography* (Boston: Houghton Mifflin, 1999).

Even after extreme: E. G. Stewart and P. Spencer, *The V Book: A Doctor's Guide to Complete Vulvovaginal Health* (New York: Bantam Books, 2002).

Around the time of: J. Jacobs Brumberg, *The Body Project: An Intimate History of American Girls* (New York: Random House, 1997): 98.

Of course, today: C. Hope, "Caucasian Female Body Hair and American Culture," *Journal of American Culture* 5, no. 1 (1982): 93.

The Brazilian: N. R. Armstrong and J. D. Wilson, "Did the 'Brazilian' Kill the Pubic Louse?" *Sexually Transmitted Infections* 82 (2006): 265–266; C. Enting, "The Brazilian Affair," *Dominion Post*, December 13, 2007.

Some Europeans actually: C. Blakemore and S. Jennett, eds., *The Oxford Companion to the Body*, (Oxford, UK: Oxford University Press, 2002): 778.

Incidentally, it's not only: For pubic hair transplants in Korea, see Y. R. Lee, S. J. Lee, J. C. Kim, and H. Ogawa, "Hair Restoration Surgery in Patients with Pubic Atrichosis or Hypotrichosis: Review of Technique and Clinical Consideration of 507 Cases," *Dermatol Surg* 32, no. 11 (2006): 1327–1335; see also C. K. Hong and H. G. Choi, "Hair Restoration Surgery in Patients with Hypotrichosis of the Pubis: The Reason and Ideas for Design," *Dermatol Surg* 25, no. 6 (1999): 475–479.

A 2006 paper: L. M. Shinmyo, F. X. Nahas, and L. M. Ferreira, "Guidelines for Pubic Hair Restoration," *Aesthetic Plastic Surgery* 30, no. 1 (2006): 104–107.

And like many rules: "Androgen Insensitivity Syndrome," *Medline Plus*, http://www.nlm.nih.gov/medlineplus/ency/article/001180.htm.

Beyond the vaginal opening: For a technical resource on female anatomy, see C. R. B. Beckmann and F. W. Ling, *Obstetrics and Gynecology*, 5th ed. (Baltimore, MD: Lippincott Williams and Wilkins, 2006).

Similarly, endurance athletes: C. L. Otis, "Exercise-Associated Amenorrhea," *Clinical Sports Medicine* 11 (1992): 351–362; M. Shangold, R. W. Rebar, A. C. Wentz, and I. Schiff, "Evaluation and Management of Menstrual Dysfunction in Athletes," *JAMA* 263 (1990): 1665–1669.

Cultural treatment of menstruation: See, for example, J. Kien, *The Battle Between the Moon and Sun: The Separation of Women's Bodies from the Cosmic Dance* (N.P.: Universal Publishers, 2003); T. Buckley and A. Gottlieb, *Blood Magic: The Anthropology of Menstruation* (Berkeley: University of California Press, 1988); S. Price, *Co-Wives and Calabashes* (Ann Arbor: University of Michigan Press, 1993); K. De Troyer, J. A. Herbert, J. A. Johnson, and A. Korte, *Wholly Woman, Holy Blood: A Feminist Critique of Purity and Impurity* (London: Continuum International Publishing Group, 2003); M. Berkowitz, "Reshaping the Laws of Family Purity for the Modern World," Committee on Jewish Law and Standards, Rabbinical Assembly, December 6, 2006; D. Smith, "Of Cycles and Scriptures: Chafing Against Rituals," *New York Times*, July 19, 2001; P. Hage and F. Harary, "Pollution Beliefs in Highland New Guinea," *Royal Anthropological Institute of Great Britain and Ireland* 16 no. 3 (September 1981): 367–375; K. O'Grady and P. Wansbrough, *Sweet Secrets: Stories of Menstruation* (Toronto: Sumach Press, 1997). P. Thomas, "Behind the Label: Tampons," *Ecologist*, November 27, 2007;

P. M. Tierno and P. M. Tierno Jr., *"The Secret Life of Germs: What They Are, Why We Need Them, and How We Can Protect Ourselves Against Them* (New York: Simon and Schuster, 2004).

According to Nancy Friedman: N. Friedman, *Everything You Must Know About Tampons* (New York: Berkley Books, 1981).

The first commercial product: S. D. Strauss, *The Big Idea: How Business Innovators Get Great Ideas to Market* (Chicago: Dearborn, 2001); A. Freeman and B. Golden, *Why Didn't I Think of That: Bizarre Origins of Ingenious Inventions We Couldn't Live Without* (New York: Wiley, 1997).

In their book: T. Heinrich and B. Batchelor, *Kotex, Kleenex, Huggies: Kimberly-Clark and the Consumer Revolution in American Business* (Columbus: Ohio State University Press, 2004).

Most women are familiar: F. G. Cunningham and J. W. Williams, *Williams Obstetrics*, 22nd ed. (New York: McGraw-Hill, 2005).

By the way, even: J. B. Becker, *Behavioral Endocrinology*, 2nd ed. (Cambridge, MA: MIT Press, 2002).

Then there's the whole: W. Cutler, "Lunar and Menstrual Phase Locking. Study of the Lunar Cycle's Influence on Menstrual Cycles," *American Journal of Obstetrics and Gynecology*, 137 (1980): 834.

There is some theorizing: D. F. Kripke, "Light Regulation of the Menstrual Cycle," in L. Wetterberg, ed., *Light and Biological Rhythms in Man* (Oxford, UK: Pergamon Press, 1993).

As the astronomer: G. O. Abell and B. Singer, *Science and the Paranormal: Probing the Existence of the Supernatural* (New York: Scribner, 1983).

We know that: A wonderful book on the subject is B. Whipple, A. Kahn Ladas, and J. D. Perry, *The G Spot and Other Discoveries About Human Sexuality* (New York: Macmillan, 2004); see also the important book by E. A. Lloyd, *The Case of the Female Orgasm: Bias in the Science of Evolution* (Cambridge, MA: Harvard University Press, 2005); L. Chambers, "The Story of O: Looking to Master the Art of Love? Try Starting with the Science," *Focus*, May 30, 2007.

More recently: B. R. Komisaruk, C. Beyer-Flores, B. Whipple, *The Science of Orgasm* (Baltimore, MD: Johns Hopkins University Press, 2006).

Researchers in Switzerland: S. Ortigue, S. T. Grafton, and F. Bianchi-Demicheli, "Correlation Between Insula Activation and Self-Reported Quality of Orgasm in Women, *Neuro Image* 37 (2007): 551–560; M. Wenner, "Sex Is Better for Women in Love: Reward Areas in the Brain Are Tied to Orgasm Quality," *Scientific American*, January 2008.

Oxytocin, called the "love hormone": M. J. Stephey, "Can Oxytocin Ease Shyness," *Time*, July 21, 2008. "What exactly happens when you have an orgasm." J. Margolis, *O: The Intimate History of the Orgasm* (New York: Grove Press, 2004).

But oxytocin doesn't just: B. K. Rothman, *In Labor: Women and Power in the Birthplace* (New York: Norton, 1981); "Level of Oxytocin in Pregnant Women Predicts Mother-Child Bond," *Science Daily*, October 16, 2007; L. F. Palmer, "Bonding Matters: The Chemistry of Attachment," *Attachment Parenting International* 5, no. 2 (2002): 2; Wenner, "Sex Is Better for Women in Love."

Dr. Kathleen C. Light: K. C. Light et al., "More Frequent Partner Hugs and Higher Oxytocin Levels Are Linked to Lower Blood Pressure and Heart Rate in Premenopausal Women," *Biol Psychol* 69, no. 1 (2005): 5–21.

Although it's still experimental: Hollander, E., Bartz, J., Chaplin, W., Phillips, A., Sumner, J., Soorya, L., et al. (2007) "Oxytocin increases retention of social cognition in autism." *Biol Psychiatry*, 61(4), 498–503.

It's also thought: M. R. Thompson, G. E. Hunt, and I. S. McGregor, "Neural Correlates of MDMA ('Ecstasy')-Induced Social Interaction in Rats," *Soc Neurosci* (2008): 1–13.

Chapter 2: Boys to Men

On top of that: J. Lever, D. Frederick, L. A. Peplau, "Does Size Matter? Men's and Women's Views on Penis Size Across the Lifespan," *Psychology of Men and Masculinity* 7, no. 3 (July 2006): 129–143; W. A. Fisher, N. R. Branscombe, C. R. Lemery, "The Bigger the Better? Arousal and Attributional Responses to Erotic Stimuli That Depict Different Size Penises," *Journal of Sex Research* 19, no. 4 (November 1983): 377–396.

A survey conducted in India: D. Grammaticus, "Condoms 'Too Big' for Indian Men," B.B.C. News Service, December 8, 2006.

So it looks: H. Wessells, T. F. Lue, and J. W. McAninch, "Penile Length in the Flaccid and Erect States: Guidelines for Penile Augmentation," *J Urol* 156, no. 3 (1996): 995–997.

Height seems to vary: Z. Awwad, M. Abu-Hijleh, S. Basri, N. Shegam, M. Murshidi, and K. Ajlouni, "Penile Measurements in Normal Adult Jordanians and in Patients with Erectile Dysfunction," *Int J Impot Res* 17, no. 2 (2005): 191–195; D. Mehraban, M. Salehi, and F. Zayeri, "Penile Size and Somatometric Parameters Among Iranian Normal Adult Men," *Int J Impot Res* 19, no. 3 (2007): 303–309; R. Ponchietti, N. Mondaini, M. Bonafe, F. Di Loro, S. Biscioni, and L. Masieri, "Penile Length and Circumference: A Study on 3,300 Young Italian Males," *Eur Urol* 39, no. 2 (2001): 183–186; K. Promodu, K. V. Shanmughadas, S. Bhat, and K. R. Nair, "Penile Length and Circumference: An Indian Study," *Int J Impot Res* 19, no. 6 (2007): 558–563.

They say that you can: J. Shah and N. Christopher, "Can Shoe Size Predict Penile Length?" *BJU International* 90 (2002): 586–587.

The normal male penis: For a thorough guide that answers most questions about men's private parts, see Y. Taguchi and M. Weisbord, *Private Parts: A Doctor's Guide to the Male Anatomy* (New

York: Doubleday, 1989). For a biography of the penis, see: D. M. Friedman, *A Mind of Its Own: A Cultural History of the Penis* (New York: Free Press, 2001).

Dr. Mark Winston: M. L. Winston, *The Biology of the Honey Bee* (Cambridge, MA: Harvard University Press, 1987).

And what happens to: Ibid.

Essentially, the foreskin: For a paper about the foreskin from a functional perspective, see: C. J. Cold and J. R. Taylor, "The Prepuce," *BJU Int* 83, suppl 1 (1999): 34–44.

Which is why some researchers: There are few issues more emotionally charged than the circumcision of young boys. For a history of the practice, see: L. B. Glick, *Marked in Your Flesh: Circumcision from Ancient Judea to Modern America* (Oxford, UK: Oxford University Press, 2005); and W. D. Dunsmuir and E. M. Gordon, "The History of Circumcision," *BJU Int* 83, suppl 1 (1999): 1–12. T. Hammond, "A Preliminary Poll of Men Circumcised in Infancy or Childhood," *BJU* Int, 83, suppl 1 (1999): 85–92. The following paper offers some background information on the foreskin (prepuce in primates): C. J. Cold and K. A. McGrath, "Anatomy and Histology of the Penile and Clitoral Prepuce in Primates," in G. C. Denniston, F. M. Hodges, and M. F. Milos, eds., *Male and Female Circumcision: Medical, Legal, and Ethical Considerations in Pediatric Practice* (New York: Springer, 1999), pp. xvi, 547.

"Circumcision has a long history: J. M. Hutson, "Circumcision: A Surgeon's Perspective," *J Med Ethics* 30, no. 3 (2004): 238–240;

Circumcision can have serious complications: Keloids can be a very serious, yet somewhat rare, complication after circumcision. See: R. Gurunluoglu, M. Bayramicli, T. Dogan, and A. Numanoglu, "Keloid After Circumcision," *Plast Reconstr Surg* 103, no. 5 (1999): 1539–1540.

In a 1999 study: K. O'Hara and J. O'Hara, "The Effect of Male Circumcision on the Sexual Enjoyment of the Female Partner," *BJU Int* 83, suppl 1 (1999): 79–84.

There are three basic types: E. Banks, O. Meirik, T. Farley, O. Akande, H. Bathija, and M. Ali, "Female Genital Mutilation and Obstetric Outcome: WHO Collaborative Prospective Study in Six African Countries," *Lancet* 367, no. 9525 (2006): 1835–1841; M. Afifi, "Female Genital Mutilation in Egypt," *Lancet* 369, no. 9576 (2007): 1858.

Or does it?: G. Kigozi et al., "The Effect of Male Circumcision on Sexual Satisfaction and Function, Results from a Randomized Trial of Male Circumcision for Human Immunodeficiency Virus Prevention, Rakai, Uganda," *BJU Int* 101, no. 1 (2008): 65–70.

Of course, the jury's: V. Marx and G. Lawton, "Circumcision: To Cut or Not to Cut?" *New Scientist*, July 16, 2008.

And all the research: For more on restoration of foreskins, see: D. Schultheiss, M. C. Truss, C. G. Stief, and U. Jonas, "Uncircumcision: A Historical Review of Preputial Restoration," *Plast Reconstr Surg* 101, no. 7 (1998): 1990–1998; and S. B. Brandes and J. W. McAninch, "Surgical Methods of Restoring the Prepuce: A Critical Review," *BJU Int* 83, suppl 1 (1999): 109–113.

For some years: A. J. Fink, "A Possible Explanation for Heterosexual Male Infection with AIDS," *N Engl J Med* 315, no. 18 (1986): 1167.

A seven-year study: S. J. Reynolds et al. "Male Circumcision and Risk of HIV-1 and Other Sexually Transmitted Infections in India," *Lancet* 363, no. 9414 (2004): 1039–1040.

Then came a pair: "Adult Male Circumcision Significantly Reduces Risk of Acquiring HIV: Trials in Kenya and Uganda Stopped Early," National Institute of Allergy and Infectious Diseases, http://www3.niaid.nih.gov/news/newsreleases/2006/AMC12_06press.htm.

And there has been: M. Garenne, "Long-Term Population Effect of Male Circumcision in Generalised HIV Epidemics in Sub-Saharan Africa," *African Journal of AIDS Research* 7, no. 1 (2008): 1–8.

J. D. Dickerman, "Circumcision in the Time of HIV: When Is There Enough Evidence to Revise the American Academy of Pediatrics' Policy on Circumcision?" *Pediatrics*, 119, no. 5 (2007): 1006–1007.

HIV isn't the only virus: B. Y. Hernandez et al., "Circumcision and Human Papillomavirus Infection in Men: A Site-Specific Comparison," *J Infect Dis* 197, no. 6 (2008): 787–794. R. H. Gray, M. J. Wawer, C. B. Polis, G. Kigozi, and D. Serwadda, "Male Circumcision and Prevention of HIV and Sexually Transmitted Infections," *Curr Infect Dis Rep* 10, no. 2 (2008): 121–127.

In an accompanying editorial: P. V. Chin-Hong, "Cutting Human Papillomavirus Infection in Men," *J Infect Dis* 197, no. 6 (2008): 781–783.

For now, I agree with: American Academy of Pediatrics' policy statement on circumcision is available at http://aappolicy.aappublications.org/cgi/content/abstract/pediatrics;103/3/686.

That theory is given: S. Shefi, P. E. Tarapore, T. J. Walsh, M. Croughan, and P. J. Turek, "Wet Heat Exposure: A Potentially Reversible Cause of Low Semen Quality in Infertile Men," *Int Braz J Urol* 33, no. 1 (2007): 50–57.

A recent study by researchers: "Hot Tubs May Cut Male Fertility," *BBC News Service*, March 5, 2007.

And a new study: A. Jung, P. Strauss, H. J. Lindner, and H. C. Schuppe, "Influence of Heating Car Seats on Scrotal Temperature," *Fertil Steril* 90, no. 2 (2008): 335–339.

During development, testicles: For more on the descent of the testicles, see W. R. Anderson, J. A. Hicks, and S. A. Holmes, "The Testis: What Did He Witness?" *BJU Int* 89, no. 9 (2002): 910–911; and L. Werdelin and A. Nilsonne, "The Evolution of the Scrotum and Testicular Descent in Mammals: A Phylogenetic View," *J Theor Biol* 196, no. 1 (1999): 61–72.

Yet having large testicles: S. Pitnick, K. E. Jones, and G. S. Wilkinson, "Mating System and Brain Size in Bats," *Proc Biol Sci* 273, no.

1587 (2006): 719–724; see also G. Vince, "Big Brain Means Small Testes, Finds Bat Study," *New Scientist*, December 7, 2005.

In 2008, scientists: J. M. Nascimento et al., "The Use of Optical Tweezers to Study Sperm Competition and Motility in Primates," *J R Soc Interface* 5, no. 20 (2008): 297–302.

According to Jaclyn Nascimento: "Primate Sperm Competition: Speed Matters," *Science Today at the University of California*, October 1, 2007, http://www.ucop.edu/sciencetoday/article/16514.

Having sticky semen: S. Dorus, P. D. Evans, G. J. Wyckoff, S. S. Choi, and B. T. Lahn, "Rate of Molecular Evolution of the Seminal Protein Gene SEMG2 Correlates with Levels of Female Promiscuity," *Nat Genet* 36, no. 12 (2004): 1326–1329; S. J. Carnahan and M. I. Jensen-Seaman, "Hominoid Seminal Protein Evolution and Ancestral Mating Behavior," *Am J Primatol* 70, no. 10 (2008): 939–948.

"It's similar to the pressures: "Sperm's Solution to Promiscuity," *BBC World News*, November 8, 2004.

Scientists have observed: U. Candolin and J. D. Reynolds, "Adjustments of Ejaculation Rates in Response to Risk of Sperm Competition in a Fish, the Bitterling (*Rhodeus sericeus*)," *Proc Biol Sci* 269, no. 1500 (2002): 1549–1553.

The followers of the Greek thinker: M. Foucault, *The History of Sexuality* (New York: Vintage Books, 1988).

Dr. Harry Fisch: N. Angier, "Sleek, Fast and Focused: The Cells That Make Dad Dad," *New York Times*, June 12, 2007.

A 2005 Australian study: S. J. Kilgallon and L. W. Simmons, "Image Content Influences Men's Semen Quality," *Biol Lett* 1, no. 3 (2005): 253–255.

Evolutionary biologist: Leigh Simmons, quoted in J. Skatssoon, "Porn Makes Sperm Better Swimmers," *ABC Science*, June 8, 2005.

But don't take my word for it: You can watch it at "The Semen Taste Test: Can food change the flavour of a man's seminal fluid?" *Sci-*

ence and Nature: The Truth About Food, http://www.bbc.co.uk/sn/humanbody/truthaboutfood/sexy/spermtaste.shtml.

Chapter 3: I'm So Excited and I Just Can't Hide It

A new study published: G. C. Gonzaga, M. G. Haselton, J. Smurda, M. S. Davies, and J. C. Poore, "Love, Desire, and the Suppression of Thoughts of Romantic Alternatives," *Evolution and Human Behavior* 29 (2008): 119–126.

In another study: Maner, J. K., Rouby, D. A., and Gonzaga, G. "Automatic Inattention to Attractive Alternatives: The Evolved Psychology of Relationship Maintenance." *Evolution & Human Behavior* 29 (2008): 343–349.

Joseph Forgas, a psychologist: R. Nowak, "Love Is Really Blind, or at Least Blinkered," *New Scientist,* July 7, 2008.

"Women know they have: Quoted in "Near Ovulation, Your Cheatin' Heart Will Tell on You," UC Newsroom, press release, http://www.universityofcalifornia.edu/news/article/7768.

A recent study led by: M. G. Haselton, M. Mortezaie, E. G. Pillsworth, A. Bleske-Rechek, and D. A. Frederick, "Ovulatory Shifts in Human Female Ornamentation: Near Ovulation, Women Dress to Impress," *Hormones and Behavior* 51, no. 1 (2007): 40–45; see also M. G. Haselton and S. W. Gangestad, "Conditional Expression of Women's Desires and Men's Mate Guarding Across the Ovulatory Cycle," *Hormones and Behavior* 49, no. 4 (2006): 509–518.

"We found that women: Dr. Martie Haselton, quoted in "Studies: Women Genetically Programmed to Cheat," *ABC News,* January 4, 2006.

A related UCLA study: E. G. Pillsworth and M. G. Haselton, "Male Sexual Attractiveness Predicts Differential Ovulatory Shifts in Female Extra-Pair Attraction and Male Mate

Retention," *Evolution and Human Behavior* 27, no. 4 (2006): 247–258.

"I hope the message women get: Elizabeth Pillsworth, quoted in "Studies: Women Genetically Programmed to Cheat."

Rachel Herz is: R. Herz, *The Scent of Desire: Discovering Our Enigmatic Sense of Smell* (New York: William Morrow, 2007).

"I knew I would marry: Estelle Campenni, quoted in E. Svoboda, "Scents and Sensibility." *Psychology Today*, January–February 2008.

"As of now: Personal interview with Charles Wysocki.

The granddaddy of research: C. Wedekind, T. Seebeck, F. Bettens, and A. J. Paepke, "MHC-Dependent Mate Preferences in Humans," *Proc Biol Sci* 260, no. 1359 (1995): 245–249.

When I interviewed Herz: Personal interview with Rachel Herz.

"There's no Brad Pitt: Rachel Herz, quoted in Svoboda, "Scents and Sensibility."

Not only does HLA compatibility: C. E. Garver-Apgar, S. W. Gangestad, R. Thornhill, R. D. Miller, and J. J. Olp, "Major Histocompatibility Complex Alleles, Sexual Responsivity, and Unfaithfulness in Romantic Couples," *Psych Sci* 17 (2006): 830–835.

Professor Devendra Singh: D. Singh and P. M. Bronstad, "Female Body Odour Is a Potential Cue to Ovulation," *Proc Biol Sci* 268, no. 1469 (2001): 797–801.

Dr. Ivanka Savic: I. Savic, H. Berglund, and P. Lindstrom, "Brain Response to Putative Pheromones in Homosexual Men," *Proc Natl Acad Sci USA* 102, no. 20 (2005): 7356–7361.

Dr. Savic followed up: H. Berglund, P. Lindstrom, and I. Savic, "Brain Response to Putative Pheromones in Lesbian Women," *Proc Natl Acad Sci USA* 103, no. 21 (2006): 8269–8274.

"Heterosexual males: Y. Martins, G. Preti, C. R. Crabtree, T. Runyan, A. A. Vainius, and C. J. Wysocki, "Preference for Human Body Odors Is Influenced by Gender and Sexual Orientation," *Psychological Science* 16 (2005): 694–701.

"Our study can't answer: Dr. Ivanka Savic, quoted in P. Barry, "Clue to Sexual Attraction Found in Lesbian Brain," *New Scientist*, May 8, 2006.

A recent study published: S. S. Young, B. Eskenazi, F. M. Marchetti, G. Block, and A. J. Wyrobek, "The Association of Folate, Zinc, and Antioxidant Intake with Sperm Aneuploidy in Healthy Non-Smoking Men," *Human Reproduction* 23, no. 5 (2008): 1014–1022.

But where does tall: N. G. Jablonski, *Skin: A Natural History* (Berkeley: University of California Press, 2006).

According to Alan Slater: Alan Slater, quoted in A. Gosline, "Babies Prefer to Gaze upon Beautiful Faces," *New Scientist*, September 6, 2004.

Slater believes babies: Ibid.

But as Charles Darwin: C. Darwin, *The Descent of Man, and Selection in Relation to Sex: The Concise Edition* (New York: Plume, 2007).

Well, in 2007 researchers: A. C. Little, C. L. Apicella, and F. W. Marlowe, "Preferences for Symmetry in Human Faces in Two Cultures: Data from the UK and the Hadza, an Isolated Group of Hunter-Gatherers," *Proc Biol Sci* 274, no. 1629 (2007): 3113–3117; A. C. Little et al., "Symmetry Is Related to Sexual Dimorphism in Faces: Data Across Culture and Species," *PLoS ONE* 3, no. 5 (2008): 2106.

Dr. Anthony Little made: Dr. Anthony Little, quoted in R. Highfield, "Symmetry Really Is Sexy, Say Scientists," *Telegraph*, November 15, 2008.

Another small but interesting: R. Thornhill, S. W. Gangestad, and R. Comer, "Human Female Orgasm and Mate Fluctuating Asymmetry," *Animal Behaviour* 50, no. 6 (1995): 1601–1615.

"We don't think women: Randy Thornhill, quoted in D. Concar, "Sex and the Symmetrical Body," *New Scientist*, April 22, 1995.

But that won't stop: Ibid.

In the masterfully compiled book: A. Taschen, ed., *Aesthetic Surgery* (London: Taschen, 2005).

Şerafeddin Sabuncuoğlu was: I. Basagaoglu, S. Karaca, and Z. Salihoglu, "Anesthesia Techniques in the Fifteenth Century by Şerafeddin Sabuncuoğlu," *Anesth Analg* 102, no. 4 (2006): 1289.

According to the American Society: For statistics on yearly cosmetic procedures in 2007, see the American Society for Aesthetic Plastic Surgery, http://www.surgery.org/press/news-release.php?iid=491.

It's not just Botox injections: A. M. Munhoz et al., "Aesthetic Labia Minora Reduction with Inferior Wedge Resection and Superior Pedicle Flap Reconstruction," *Plast Reconstr Surg* 118, no. 5 (2006): 1237–1250.

Perhaps when girls as young: A. Lynch, M. Marulaiah, and U. Samarakkody, "Reduction Labioplasty in Adolescents," *J Pediatr Adolesc Gynecol* 21, no. 3 (2008): 147–149.

An Australian study says: G. Rhodes, L. W. Simmons, and M. Peters, "Attractiveness and Sexual Behavior: Does Attractiveness Enhance Mating Success?" *Evolution and Human Behavior* 26 (2005): 186–201.

And a British study: K. K. Kampe, C. D. Frith, R. J. Dolan, and U. Frith, "Reward Value of Attractiveness and Gaze," *Nature* 413, no. 6856 (2001): 589.

"Meeting a potential good friend: Knut Kampe, quoted in H. Muir, "Beautiful People Spark a Brain Reaction," *New Scientist*, October 10, 2001.

There has long been: D. Blum, *Sex on the Brain: The Biological Differences Between Men and Women* (New York: Viking, 1997).

"Most men are very visual: Dr. Daniel Amen, quoted in "Use Your Brain to Have Hotter Sex," *Today*, http://www.msnbc.msn.com/id/16673923/page/2/.

And there have certainly: S. Hamann, R. A. Herman, C. L. Nolan, and K. Wallen, "Men and Women Differ in Amygdala Response to Visual Sexual Stimuli," *Nat Neurosci* 7, no. 4 (2004): 411–416.

Another 2004 study: M. L. Chivers, G. Rieger, E. Latty, and J. M. Bailey, "A Sex Difference in the Specificity of Sexual Arousal," *Psychol Sci* 15, no. 11 (2004): 736–744.

Then, in 2007, researchers: T. M. Kukkonen, Y. M. Binik, R. Amsel, and S. Carrier, "Thermography as a Physiological Measure of Sexual Arousal in Both Men and Women," *Journal of Sexual Medicine* 4 (2007): 93–105.

Also in 2007, Kim: H. A. Rupp and K. Wallen, "Sex Differences in Viewing Sexual Stimuli: An Eye-Tracking Study in Men and Women," *Hormones and Behavior* 51, no. 4 (2007): 524–533.

Surprisingly, Rupp said: Dr. Heather Rupp, quoted in "A Selection of Kinsey Institute Research," http://www.indiana.edu/~kinsey/research/summary-photos.html.

By the way, paying: R. O. Deaner, A. V. Khera, and M. L. Platt, "Monkeys Pay Per View: Adaptive Valuation of Social Images by Rhesus Macaques," *Curr Biol* 15, no. 6 (2005): 543–548.

"Flirting is a way of testing: Arthur Aron, quoted in B. Luscombe, "The Science of Romance: Why We Flirt," *Time*, 2008.

Regardless of how: Charles Darwin, *The Expression of the Emotions in Man and Animals* (Chicago: University of Chicago Press): 372.

In fact, a new study: S. M. Hughes, M. A. Harrison, and G. G. Gallup Jr., "Sex Differences in Romantic Kissing Among College Students: An Evolutionary Perspective," *Evolutionary Psychology* 5, no. 3 (2007): 612–631.

In an article in: C. Walter, "Affairs of the Lips," *Scientific American*, January 2008.

Chapter 4: Let's Talk About Sex

Among the Yungar people: A. J. Hobday, L. Haury, and P. K. Dayton, "Function of the Human Hymen," *Med Hypotheses* 49, no. 2 (1997): 171–173.

According to Jelto Drenth: J. Drenth, *The Origin of the World: Science and Fiction of the Vagina* (London: Reaktion Books, 2005).

The *Trotula*, a medieval medical text: M. H. Green, *The Trotula: An English Translation of the Medieval Compendium of Women's Medicine* (Philadelphia: University of Pennsylvania Press, 2002).

These sham virginity procedures: E. Sciolina and S. Mekhennet, "In Europe, Debate over Islam and Virginity," *New York Times,* June 11, 2008; B. Crumley, "The Dilemma of Virginity Restoration," *Time,* July 13, 2008.

According to the *Kama Sutra*: A. Daniélou, *The Complete Kama Sutra: The First Unabridged Modern Translation of the Classic Indian Text* (South Paris, ME: Park Street Press, 1994).

A study from the *Archives*: J. A. Adams, A. S. Botash, and N. Kellogg, "Differences in Hymenal Morphology Between Adolescent Girls With and Without a History of Consensual Sexual Intercourse," *Arch Pediatr Adolesc Med* 158, no. 3 (2004): 280–285; Adams, J. A. "Guidelines for medical are of children evaluated for suspected sexual abuse: an update for 2008" *Curr Opin Obstet Gynecol,* 20(5), (2008) 435–441. McCann, J., Miyamoto, S., Boyle, C., and Rogers, K., (2007) "Healing of hymenal injuries in prepubertal and adolescent girls: a descriptive study." *Pediatrics,* 119(5), e1094–1106. D. Holtzman and N. Kulish, "Nevermore: The Hymen and the Loss of Virginity," *J Am Psychoanalytic Assoc* 47 (1999): 1461–1464.

Oxytocin and vasopressin: Hammock, E. A. "Gene regulation as a modulator of social preference in voles." *Adv Genet,* 59, (2007) 107–127.

Fascinating research: Walum, H., Westberg, L., Henningsson, S., Neiderhiser, J. M., Reiss, D., Igl, W., et al. "Genetic variation in the vasopressin receptor 1a gene (AVPR1A) associates with pair-bonding behavior in humans." *Proc Natl Acad Sci U S A,* 105(37) (2008): 14153–14156.

"Men with two: H. Walum, quoted in S. Vedantam, "Study Links Gene Variant in Men to Marital Discord," *Washington Post*, September 2, 2008.

"There are many: H. E. Fisher, *Why We Love: The Nature and Chemistry of Romantic Love* (New York: Holt, 2004).

"An intricately coordinated: B. R. Komisaruk, C. Beyer, and B. Whipple, *The Science of Orgasm* (Baltimore, MD: Johns Hopkins University Press, 2006).

But as Dr. Elisabeth Lloyd: E. A. Lloyd, *The Case of the Female Orgasm: Bias in the Science of Evolution* (Cambridge, MA: Harvard University Press, 2005).

Baker said the images: Robin Baker, quoted in Michael Wedeman, "Sperm Wars: The Science of Sex," *Journal of Sex Research*, reprinted at http://findarticles.com/p/articles/mi_m2372/is_/ai_20746731.

Lloyd, in fact: E. A. Lloyd, The Case of the Female Orgasm: Bias in the Science of Evolution (Cambridge, MA: Harvard University Press,2005); D. Symons, *The Evolution of Human Sexuality* (New York: Oxford University Press, 1979).

"Females Get the Nerve Pathways: Elizabeth Lloyd, quoted in D. Smith, "A Critic Takes On the Logic of the Female Orgasm," *New York Times*, May 17, 2005.

"Recreational sex: J. M. Diamond, *Why Is Sex Fun? The Evolution of Human Sexuality* (New York: HarperCollins, 1997).

With all the attention focused: A. F. Kiroglu, H. Bayrakli, K. Yuca, H. Cankaya, and M. Kiris, "Nasal Obstruction as a Common Side-Effect of Sildenafil Citrate," *Tohoku Journal of Experimental Medicine* 208, no. 3 (2006): 251–254.

In his treatise: *Hippocratic Writings*, translated by J. Chadwick, W. N. Mann, E. T. Withington, and I. M. Lonie. (New York: Penguin Classics, 1984).

The definitive Sanskrit text: Daniélou, *The Complete Kama Sutra*. Dhall, "Adolescence: Myths and Misconceptions."

Even twentieth-century American: R. Chalker, *The Clitoral Truth: The Secret World at Your Fingertips* (New York: Seven Stories, 2002).

Maxine Davis, author of: Davis, M. *Sexual Responsibility in Marriage.* (New York: Dial Press, 1963).

"No matter how tired: C. B. S. Evans, N. Haire, and R. L. Dickinson, *Sex Practice in Marriage* (New York: Emerson, 1935).

As Dr. Whipple: F. Addiego, E. G. Belzer, J. Comolli, W. Moger, J. D. Perry, and B. Whipple, "Female Ejaculation: A Case Study," *Journal of Sex Research* 17 (1981): 13–21.

In a 1988 study: M. Zaviačič, A. Zaviačičova, I. K. Holomáň, J. Molčan, "Female Urethral Expulsions Evoked By Local Digital Stimulation of the G-Spot: Differences in the Response Patterns," *Journal of Sex Research* 24 (1988): 311–318.

That study was followed by: C. A. Darling, J. K. Davidson Sr., and C. Conway-Welch, "Female Ejaculation: Perceived Origins, the Grafenberg Spot/Area, and Sexual Responsiveness," *Arch Sex Behav* 19, no. 1 (1990): 29–47.

No less a sexual science celebrity: E. Grafenberg, "The Role of Urethra in Female Orgasm," *International Journal of Sexology* 3 (1950): 144–145.

In 2002, the British Board of Film: B. Arndt, "Women and the UFO of Sex," *The Sydney Morning Herald,* April 19, 2003.

Philadelphia gynecologist: A. K. Ladas, B. Whipple, and J. D. Perry, *The G Spot and Other Recent Discoveries About Human Sexuality* (New York: Holt, Rinehart, and Winston, 1982).

The vulva and vagina were: Ibid.

The most recent study of: F. Wimpissinger, K. Stifter, W. Grin, and W. Stackl, "The Female Prostate Revisited: Perineal Ultrasound and Biochemical Studies of Female Ejaculate," *Journal of Sex Medicine* 4, no. 5 (2007): 1388–1393.

Sex educator Deborah Sundahl: D. Sundahl, *Female Ejaculation and the G-Spot* (Alameda, CA: Hunter House, 2003).

One thing is sure: Wimpissinger et al., "The Female Prostate Revisited."

When I spoke: Personal interview with Beverly Whipple.

"Begin with clitoral: Ibid.

As one sufferer said to: Judith Reichman, "Help! What Do I Do About 'Honeymoon Cystitis'?" *Today*, http://www.msnbc.msn .com/id/9225687/.

"Obviously, it is difficult: N. B. McCormick, "When Pleasure Causes Pain: Living with Interstitial Cystitis," *Sexuality and Disability* 17 (2004): 7–18.

"Whipple and Perry hypothesized: Ladas, A. K., Whipple, B., and Perry, J. D. *The G Spot and Other Recent Discoveries About Human Sexuality*. (New York: Holt, Rinehart, and Winston, 1982).

Professor Itzhak Ofek: See the press release: www. eurekalert.org/ pub_releases/2008–01/afot-cra011008.php; Ofek, I., Goldhar, J., Zafriri, D., Lis, H., Adar, R., and Sharon, N. "Anti-Escherichia coli adheson activity of cranberry and blueberry juices." (1991) *N Engl J Med*, 324(22), 1599.

But it may be: S. A. Robertson, W. V. Ingman, S. O'Leary, D. J. Sharkey, and K. P. Tremellen. "Transforming Growth Factor Beta—A Mediator of Immune Deviation in Seminal Plasma," *J Reprod Immunol* 57, no. 1–2 (2002): 109–128.

This theory is supported: R. B. Ness, D. A. Grainger, "Male Reproductive Proteins and Outcomes," *Am J Obstet Gynecol*, 198, 6 (2008): 620el–4; C. A. Koelman, A. B. Coumans, H. W. Nijman, II Doxiadis, G. A. Dekker, and F. H. Claas, "Correlation Between Oral Sex and a Low Incidence of Preeclampsia: A Role for Soluble HLA in Seminal Fluid?" *J Reprod Immunol* 46, no. 2 (2000): 155–166.

Chapter 5: Come as You Are

One of the best explications: M. Ridley, *The Red Queen: Sex and the Evolution of Human Nature* (New York: Perennial, 2003).

Cheryl Chase, executive director: Sally Lehrman, "Going beyond X and Y," *Scientific American*, May 2007.

In December 2006, Santhi: Katie Law, "The Indelicate Art of Gender Testing," *Inkling Magazine*, January 17, 2007, http://www.inklingmagazine.com/articles/gender-testing/.

The magazine *New Scientist*: A. George, "Teenagers Trapped in the Wrong Body," *New Scientist*, 2007.

So what is GID?: Ibid.

A new study: E. K. Bentz, L. A. Hefler, U. Kaufmann, J. C. Huber, A. Kolbus, and C. B. Tempfer, "A Polymorphism of the CYP17 Gene Related to Sex Steroid Metabolism Is Associated with Female-to-Male But Not Male-to-Female Transsexualism," *Fertil Steril* 90, no. 1 (2008): 56–59.

Dr. Eric Vilain: S. Lehrman, "Going beyond X and Y," *Scientific American*, May 20, 2007.

Chapter 6: Let It Be

In 1972, Linda Wolfe: Personal interview.

Evolutionary biologist Paul Vasey: P. L. Vasey, "Function and Phylogeny: The Evolution of Same-Sex Sexual Behavior in Primates," *Journal of Psychology and Human Sexuality* 18 (2007): 215–244.; P. L. Vasey, "Sex Differences in Sexual Partner Acquisition, Retention, and Harassment During Female Homosexual Consortships in Japanese Macaques," *American Journal of Primatology* 64 (2004): 397–409;

As Vasey succinctly states: Paul Vasey, quoted in Stephanic Chasteen, "Gender Scientists Explore a Revolution in Evolution," *Stanford Report*, February 19, 2003, http://news-service.stanford.edu/news/2003/february19/aaassocialselection219.html.

When I asked her: Personal interview with Linda Wolfe.

Macacques are certainly not: J. Roughgarden, *Evolution's Rainbow: Diversity, Gender, and Sexuality in Nature and People* (Berkeley: University of California Press, 2004).

My discipline teaches: Joan Roughgarden, quoted in J. Lehrer, "The Gay Animal Kingdom," *Seed Magazine*, June 2006, http://seed-magazine.com/news/2006/06/the_gay_animal_kingdom. php.

Male big horn sheep: Ibid.

Cataloging animal homosexuality: B. Bagemihl, *Biological Exuberance: Animal Homosexuality and Natural Diversity* (New York: St. Martin's Press, 1999).

And, if you have any doubt: C. W. Moeliker, "The First Case of Homosexual Necrophilia in the Mallard *Anas Platyrhynchos* (Aves: Anatidae)," *DEINSEA* 8 (2001): 243–247.

"If you're looking for: Robin Dunbar, quoted in G. Vines, "Queer Creatures," *New Scientist*, August 7, 1999.

"One plausible explanation: Ibid.

"Same-sex sexuality: Ibid.

"Although the first reports: Ibid.

"As biologist Valerius Geist: Ibid.

When it comes to humans: K. J. Verweij et al., "Genetic and Environmental Influences on Individual Differences in Attitudes Toward Homosexuality: An Australian Twin Study," *Behav Genet* 38, no. 3 (2008): 257–265.

One of the most prominent: J. M. Bailey and R. C. Pillard, "A Genetic Study of Male Sexual Orientation," *Arch Gen Psych* 48, no. 12 (1991): 1089–1096.

With the research: Simon LeVay, quoted in Neil Swidey, "What Makes People Gay?" *Brain Research Institute*, August 12, 2005, http://www.bri.ucla.edu/bri_weekly/news_050812.asp.

Dr. Camperio-Ciani's latest study: A. Camperio-Ciani, F. Iemmola and S. R. Blecher, "Genetic Factors Increase Fecundity in Female Material Relatives of Bisexual Men as in Homosexuals," *J Sex Med 6*, 2 (2008): 449–455; F. Iemmola, A. Camperio-Ciani, "New Evidence of Genetic Factors Influencing Sexual Orientation in Men: Female Fecundity Increase in the Maternal Line," *Arch Sex Behavior* [Epub ahead of print]; A. Camperio-Ciani, F.

Corna, and C. Capiluppi, "Evidence For Maternally Inherited Factors Favouring Male Homosexuality and Promoting Female Fecundity," *Proc Biol Sci* 271, no. 1554 (2004): 2217–2221.

"We have finally solved: Andrea Camperio-Ciani, quoted in A. Coghlan, "Survival of Genetic Homosexual Traits Explained," *New Scientist*, October 13, 2004.

"We think of it: Simon LeVay, quoted in Ibid.

"It helps to answer: Dean Hamer, quoted in T. Osborne, "Bisexuality Passed On by 'Hyper-Heterosexuals,'" *New Scientist*, August 15, 2008.

"Our findings are only one piece: Camperio-Ciani, quoted in Coghlan, "Survival of Genetic Homosexual Traits Explained."

"Genetics is not determining: Andrea Camperio-Ciani, quoted in "How Bisexuality Is Passed On in the Genes," *New Scientist*, August 20, 2008.

"I haven't found one:" William Reiner, quoted in N. Swidey, "What Makes People Gay?" *Boston Globe*, August 14, 2005.

"Exposure to male hormones: Ibid.

More evidence that prenatal hormone: T. J. Williams et al., "Finger-Length Ratios and Sexual Orientation," *Nature* 404, no. 6777 (2000): 455–456.

Chapter 7: Tainted Love

A team of researchers: A. K. Chaturvedi, E. A. Engels, W. F. Anderson, and M. L. Gillison, "Incidence Trends for Human Papillomavirus-Related and -Unrelated Oral Squamous Cell Carcinomas in the United States," *J Clin Oncol* 26, no. 4 (2008): 612–619.

"What we do know: Lesley Walker, quoted in M. Day, "Oral Sex-Related Cancer at 30-Year High," *New Scientist*, 2008.

"We need to start: Dr. Maura Gillison, quoted in "HPV Increasingly Causes Oral Cancer in Men," Associated Press, February 1, 2008, http://www.msnbc.msn.com/id/22956090/.

She believes there's: M. L. Gillison, "Human Papillomavirus-Related Diseases: Oropharynx Cancers and Potential Implications for Adolescent HPV Vaccination," *J Adolesc Health* 43, no. 4 suppl (2008): S52–S60. For background on cervical cancer survival, see: A. K. Chaturvedi et al., "Second Cancers Among 104,760 Survivors of Cervical Cancer: Evaluation of Long-Term Risk," *J Natl Cancer Inst* 99, no. 21 (2007): 1634–1643.

And a recent study: A. B. James, T. Y. Simpson, and W. A. Chamberlain, "Chlamydia Prevalence Among College Students: Reproductive and Public Health Implications," *Sex Transm Dis* 35, no. 6 (2008): 529–532. For more information on Gonorrhea, see: A. L. Garcia, V. K. Madkan, and S. K. Tyring, "Gonorrhea and Other Venereal Diseases," in K. Wolff et al., eds., *Fitzpatrick's Dermatology in General Medicine*, 7th ed. (New York: McGraw-Hill, 2008).

In 2004, a team of scientists: A. Idahl, J. Boman, U. Kumlin, and J. I. Olofsson, "Demonstration of Chlamydia Trachomatis Igg Antibodies in the Male Partner of the Infertile Couple Is Correlated with a Reduced Likelihood of Achieving Pregnancy," *Hum Reprod* 19, no. 5 (2004): 1121–1126.

In 2007 José Luis Fernández: G. Gallegos, B. Ramos, R. Santiso, V. Goyanes, J. Gosalvez, and J. L. Fernández, "Sperm DNA Fragmentation in Infertile Men with Genitourinary Infection by Chlamydia Trachomatis and Mycoplasma," *Fertil Steril* 90, no. 2 (2008): 328–334. For STI background, see: K. J. Ryan, C. G. Ray, and J. C. Sherris, *Sherris Medical Microbiology: An Introduction to Infectious Diseases*, 4th ed. (New York: McGraw-Hill, 2004).

Damage from herpes: S. Jha and R. Patel, "Klüver-Bucy Syndrome: An Experience with Six Cases," *Neurol India* 52, no. 3 (2004): 369–371; S. Pradhan, M. N. Singh, and N. Pandey, "Kluver Bucy Syndrome in Young Children," *Clin Neurol Neurosurg* 100, no. 4 (1998): 254–258.

New York psychiatrist: L. R. Tancredi, *Hardwired Behavior: What*

Neuroscience Reveals About Morality (New York: Cambridge University Press, 2005).

A description of a syphilis sore: M. Zuk, "A Great Pox's Greatest Feat: Staying Alive," *New York Times*, April 29, 2008.

Medical historian Deborah Hayden: D. Hayden, *Pox: Genius, Madness, and the Mysteries of Syphilis*. (New York: Basic Books, 2003).

But how did a disease: R. J. Knell, "Syphilis in Renaissance Europe: Rapid Evolution of an Introduced Sexually Transmitted Disease?" *Proc Biol Sci* 271, suppl 4 (2004): S174–S176.

A sweeping survey: A. M. Houston, J. Fang, C. Husman, and L. Peralta, "More Than Just Vaginal Intercourse: Anal Intercourse and Condom Use Patterns in the Context of 'Main' and 'Casual' Sexual Relationships Among Urban Minority Adolescent Females," *J Pediatr Adolesc Gynecol* 20, no. 5 (2007): 299–304.

Now, a new study conducted by: A. A. Adimora, V. J. Schoenbach, and I. A. Doherty, "Concurrent Sexual Partnerships Among Men in the United States," *Am J Public Health* 97 (2007): 2230–2237.

"This Study Sheds Light: Adaora A. Adimora, quoted in Amy Norton, "Concurrent Sex Partners Not Uncommon for U.S. Men," Reuters, October 30, 2007, http://uk.reuters.com/article/health-News/idUKSAT07904620071030.

"People—especially women: Ibid.

As Mullins points out: Jolene Mullins, quoted in Liz Langley, "Sex and the Single Septuagenarian," *Salon*, December 4, 2006, http://www.salon.com/mwt/feature/2006/12/04/senior_std/print.html.

Chapter 8: Jagged Little Pill

James Higham, a research fellow: J. P. Higham, C. Ross, Y. Warren, M. Heistermann, and A. M. Maclarnon, "Reduced Reproductive Function in Wild Baboons (*Papio Hamadryas Anubis*) Related to Natural Consumption of the African Black Plum (*Vitex Doniana*)," *Hormones and Behavior* 52, no. 3 (2007): 384–390.

The plum "appears to: Ibid.

But one primate expert: Patricia Whitten quoted in "Primates on the Pill," *New Scientist*, 2007.

One controversial theory: H. M. Bruce, "An Exteroceptive Block to Pregnancy in the Mouse," *Nature* 184 (1959): 105.

A study by the World: World Health Organization, "The World Health Organization Multinational Study of Breast-Feeding and Lactational Amenorrhea. III. Pregnancy During Breast-Feeding," *Fertility and Sterility* 72 (1999): 431–440.

A recent study led by: M. Emery-Thompson et al., "Aging and Fertility Patterns in Wild Chimpanzees Provide Insights into the Evolution of Menopause," *Current Biology* 17 (2007): 2150–2156.

"Females in the wild: Melissa Emery-Thompson, quoted in R. Hooper, "Menopause Sets Humans Apart from Chimps," *New Scientist*, 2007.

"Human life history: Ibid.

A study led by Daryl P. Shanley: D. P. Shanley, R. Sear, R. Mace, and T. B. Kirkwood, "Testing Evolutionary Theories of Menopause," *Proc Biol Sci* 274, no. 1628 (2007): 2943–2949.

"Our results point clearly: R. Highfield, "How the Menopause Allows Granny to Help," *Telegraph*, 2007.

In *Contraception and Abortion*: J. M. Riddle, *Contraception and Abortion from the Ancient World to the Renaissance* (Cambridge, MA: Harvard University Press, 1992).

A suggestion is made: Ibid.

It was not long: Ibid.

Casanova wore the finest: Ibid.

Not everyone credits: Kelly, I. *Casanova: Actor Lover Priest Spy*. (New York: Jeremy P. Tarcher/Penguin, 2008). A. Tone, et al., *Devices and Desires: A History of Contraceptives in America* (New York: Hill and Wang, 2001).

"In addition to being responsible: A. Tone, etc.

It is a vicious cycle: Handbill advertising Sanger's first clinic in Brooklyn, NY.

The campaign for birth control: Ibid.

It was one of the greatest: Ibid.

The Pill has long been rumored: S. Gupta, "Weight Gain on the Combined Pill—Is It Real?" *Hum Reprod Update* 6, no. 5 (2001): 427–431.

There is also some evidence: J. Kulkarni, "Depression as a Side Effect of the Contraceptive Pill," *Expert Opin Drug Saf* 6, no. 4 (2007): 371–374.

A very recent study published: S. C. Roberts, L. M. Gosling, V. Carter, and M. Petrie, "MHC-Correlated Odour Preferences in Humans and the Use of Oral Contraceptives," *Proc Biol Sci*, 275, no. 1652 (2008): 2715–2722.

Dr. Dustin Penn, director: Quotes in A. Motluk, "Scent of a Man," *New Scientist,* February 10, 2001.

According to Roberts and: Roberts et al., "MHC-Correlated Odour Preferences."

And psychologist Rachel Herz: Personal interview with Rachel Herz.

When it comes to women: Ibid.

Once you've fallen in love: Ibid.

If I now smell you: Ibid.

Recent research published: J. E. Chavarro, T. L. Toth, S. M. Sadio, and R. Hauser, "Soy Food and Isoflavone Intake in Relation to Semen Quality Parameters Among Men from an Infertility Clinic," *Human Reproduction* 23, no. 11 (2008): 2584–2590.

Chapter 9: Good Vibrations

Psychologist Geoffrey Miller: G. F. Miller, *The Mating Mind: How Sexual Choice Shaped the Evolution of Human Nature* (New York: Doubleday, 2008).

acknowledgments

To Claire Wachtel, whose keen insight, infinite patience and belief in this project helped make this book possible. To Jonathan Burnham, Kathy Schneider, Christine Boyd, Julia Novitch, and Kevin Callahan, as well as the rest of the folks at Harper, whose dedication and accommodation greatly enhanced the process of writing this book. I am very grateful to all the people who gave up their time to work so relentlessly to improve the manuscript. To my publicist, Katherine Beitner, for constantly infecting others with her excitement about this project. To the always resourceful and reliable Roxanne Khamsi whose assistance was indispensable. Thanks to Richard Verver who helped with fact checking. I must also thank the incredible anatomical duo of Drs. Jeffrey T. Laitman and Joy S. Reidenberg, for their diligence and attention to detail. And to Catherine Delphia, whose talent and endless commitment to this project was greatly appreciated. My trusty agent, Jennifer Joel from ICM, was indispensable from the start. And to El, my parents, my sisters, their husbands, and their kids who are always quick to love. And of course to Shira, who sweetens my life as well the lives of those around her immeasurably.

index

9 780061 479663